Praise for

METAMORPHOSIS

"An absorbing and beautifully written exploration of biological transformation and generation, winding through the startling history of this field, accompanied by an array of fascinating characters, and with a moving personal dimension."

—Peter Godfrey-Smith, author of *Other Minds*

"Oren Harman's *Metamorphosis* is a soulful, genre-defying inquiry into the nature of transformation. Harman interweaves a history of scientific discovery with philosophy and memoir, introducing readers to astonishing characters and discoveries that bring us from the depths of the ocean to the furthest reaches of outer space. *Metamorphosis* asks difficult questions with great tenderness and deep humanity. It is a book full of wonder and revelation."

—Lauren Redniss, author of *Radioactive*

"A masterful tale of the long quest to understand one of the most wondrous and enigmatic phenomena in the animal world. Fueled by Oren Harman's boundless curiosity and rich storytelling, *Metamorphosis* roams the globe to meet fascinating creatures and equally colorful naturalists determined to penetrate their secrets. A thoroughly enjoyable and illuminating journey."

—Sean B. Carroll, author of *The Serengeti Rules*

"Inspired by history, language, and biology, Oren Harmen explores a wonderland of animal lifecycles to tell a truly fascinating tale of transformations in body and identity. His story culminates in the growth of a human child and asks how is it possible to remain ourselves while changing all the time? A book to treasure."

—Janet Browne, Harvard University, author of *Charles Darwin: A Biography*

"Stranger than the strangest imaginings of ancient mythology or science fiction are the metamorphoses undergone by the most unprepossessing of organisms: the immortal medusa, the starfish that is simultaneously child and adult, the axolotl in which the parent is the child to its offspring. Oren Harman weaves together science, history, philosophy, and the musings of a parent-to-be in this beautiful book about the twists and turns in the plot of life."

—Lorraine Daston, author of *Objectivity*

METAMORPHOSIS

METAMORPHOSIS

A NATURAL AND HUMAN HISTORY

OREN HARMAN

BASIC BOOKS
New York

Cover design by Chin-Yee Lai
Cover images © kamomeen / Shutterstock.com;
© Natural History Museum, London / Bridgeman Images
Cover copyright © 2025 by Hachette Book Group, Inc.

Basic Books
Hachette Book Group
1290 Avenue of the Americas, New York, NY 10104
www.basicbooks.com

Printed in the United States of America

First Edition: October 2025

Published by Basic Books, an imprint of Hachette Book Group, Inc. The Basic Books name and logo is a registered trademark of the Hachette Book Group.

Print book interior design by Bart Dawson.

Library of Congress Control Number: 2025006249

ISBNs: 9781541607606 (hardcover), 9781541607590 (ebook)

LSC-C

Printing 1, 2025

For Sol

CONTENTS

PART ONE
WHERE DO WE COME FROM?

PART TWO
WHERE ARE WE GOING?

PART THREE
WHAT IS THE SELF?

METAMORPHOSIS

PREFACE

For me it began with *The Very Hungry Caterpillar.* My mother would turn its thick pages ever so slowly, seducing my imagination. What will come next? It seemed like make-believe. *This* turning into *that*?? It made no sense.

Later, like many other kids, I kept a "pet" caterpillar of my own, diligently feeding it with plants and watching it grow. One morning, sure enough, it disappeared into a golden chrysalis. Holding my breath, I'd stay awake into the small hours of the night with a flashlight strapped to my forehead and a little black notebook to scribble down observations. A sparkling blue butterfly appeared one afternoon, casually perched on a leaf as if nothing special had happened. But my heart was racing. The tips of my ears went warm. And I laughed uncontrollably with the purest feelings of joy.

Forty years passed. I traveled, wrote books, fell in love, became a parent. On special nights, my mother would read *The Very Hungry Caterpillar* to my own two little kids. Then one morning, a good number of years after my second child's birth, my wife emerged from the shower holding a white stick with two pink lines on it. And suddenly, night after night, the blue butterfly from my childhood aquarium came back to me in my dreams.

Whether or not you've had kids before, to become a parent in late middle age is special. It makes you newly aware of the so-called cycle of life. More and more, I began to feel that it was getting hard to separate my personal and professional lives. I had spent a career thinking about the ideas of Charles Darwin, and the first thing my teachers taught me was that with no foresight, or plan of action, life drifts like a speck of pollen in the wind. And yet, every life-form has a beginning, an end, and a middle. An acorn becomes an oak tree, a tiny tadpole a leaping frog. Are our human lives also a kind of unfurling, a kind of working out of a preordained path?

An amalgam of two Greek words, *metamorphosis* means "transformation" or "transforming." But the biological definition of *metamorphosis*—radical post-embryonic development—is actually arbitrary, a matter of convention. How "radical" does change have to be to be considered "metamorphic," and who gets to decide? We may not turn from caterpillars into butterflies, but most of us would agree that we undergo pretty dramatic post-embryonic development. Creatures are said to be either metamorphic or not, but in reality they lie on a continuum. Having evolved from an ancient progenitor of jellyfish, in our own special way, we too metamorphose.

There was little chance of our not noticing. Moons sliver, seeds sprout, empires rise and crumble. Everything in the world around us, including our bodies, is in flux. And yet as witnesses to change, we have been and remain ambivalent: We cheer the flowering bud but mourn the rotting fruit. We search for our younger selves in mirrors while praising the wisdom that comes with wrinkles. Carried by hopes and shackled by memories, we struggle to live in the moment. Perhaps this is why in so many cultures, we've imagined ourselves with the following metaphors: earth molded and remolded, flowing rivers, passing shadows, falling leaves. Many things change us: ideas, travels, dreams. But nothing more than the people who depart from us, and the ones we bring to life.

"How many creatures walking on this earth / Have their first being in another form?" the Roman poet Ovid asked two thousand years ago. He could not have known the full extent of the truth. That creatures transform ran counter to Aristotelian science. It was considered a heresy by many even in Darwin's day. Yet according to current estimates, nearly three-quarters of all animal species on Earth undergo a form of metamorphosis.

Consider the spadefoot toad.

Named for the spades on its hindfeet that it uses to dig itself into the soil of the arid Sonoran Desert, this toad, which is really a frog,* is a burrowing animal, and it bleats like a goat. Every summer, after weeks of intensive feeding, it disappears underground. It will remain there, as still as a rock, protected by the earth's moisture. It's emergence, after ten whole months, was the inspiration for what to my mind is one of the most perfect descriptions in the English language, penned by George Orwell:

> Something—some kind of shudder in the earth, or perhaps merely a rise of a few degrees in the temperature—has told him it is time to wake up. . . . At this period, after his long fast, the toad has a very spiritual look, like a strict Anglo-Catholic towards the end of Lent. His movements are languid but purposeful, his body is shrunken, and by contrast his eyes look abnormally large. This allows one to notice, what one might not at any other time, that a toad has about the most beautiful eye of any living creature. It is like gold, or more exactly it is like the golden-coloured semi-precious stone which one sometimes sees in signet rings, and which I think is called chrysoberyl.

It is enough to take one's breath away. And yet the really astonishing drama of the toad's life has already taken place. In the process of turning from a tadpole into an adult, the spadefoot toad switched from breathing underwater to breathing air on land. It exchanged

* Generally speaking, frogs have long legs and smooth skins covered in mucus, whereas toads have shorter legs and rougher, thicker skins. Toads generally lay their eggs in long strands, whereas frogs lay theirs in a cluster, like grapes or tapioca. Most frogs have vertical pupils, whereas pupils of true toads are horizontal. But for everything just mentioned, there are also exceptions.

aquatic for terrestrial locomotion, a diet of plants and tiny water bugs for flies, crickets, and earthworms. The toad's skin had to adapt to life above water, and to the sun and winds and rains. Gills became lungs. It grew a new cardiovascular system, to carry hemoglobin, and a new immune system, since dangers that lurk underwater are quite different from those that prowl on land. New color patterns invaded the skin, to make terrestrial camouflage possible. The throat was recut, in sync with new chest muscles, to allow females and males to breathe above water, but also to call each other to mate. Front and back legs grew internally, then popped out spontaneously. The tail was absorbed into the body, the gut was shortened, heart arteries newly tooled, the eyes reformed into the gems that enchanted Orwell. A tiny tadpole brain was rewired and encased in a bony cranium. There was massive destruction of tissues and organs. Cell death. Regeneration. Genetic reprogramming. Anatomical remolding. And all this in a matter of weeks.

Metamorphosis is wild. And it poses a tough evolutionary question: Why in the world all this wasted energy and time? Prick a pupa with a pin and watch an oozing sludge trickle down; the entombed caterpillar has dissolved into goo in order to rebuild itself virtually from scratch. To emerge a butterfly, a brand new brain and wings and eyes and legs and mouthparts will have to be constructed. What problem is nature trying to solve by using this excruciating, inelegant mechanism? Why has nature invented something that makes life-forms so vulnerable? The whole thing seems mad.

As I imagined our baby developing, I knew she, too, would undergo dramatic changes. The first cleavage into two, four, eight—and by day four, sixteen cells, a microscopic speck. A neural tube forming. A tiny heartbeat appearing, bones and little fingers emerging, as if from thin air. In the coming weeks the embryo would begin to gain the features of a human, turning into a fetus and beginning to blink. Brain cells would wire like crisscrossing speed skaters at a dizzying clip. And then, when the time was right, a marvel: Having exchanged gases in the womb through my wife Yaeli's placenta, our baby would expel the water inside her lungs, and breathe air for the very first time. Slowly, over the fourth trimester, she'd begin to focus her eyes. And one day she'd look at us and smile.

The blue butterfly appeared in my dreams for a reason. Few experiences are as life-changing as bringing a kid to the world, and fewer wonders more exhilarating than the natural magic of metamorphosis. Both mark beginnings, but are, in fact, continuations. Both, in different ways, are also forms of endings. And both make us wonder about the riddles of our world.

And so I decided to write a book about metamorphosis. Not a biological textbook, or a philosophical treatise, or even a straightforward history of science. It would be a meditation, of a father-to-be. I was curious to uncover the scientific story of metamorphosis, which I was astonished to learn had not really been told. But again and again, the human themes kept emerging. The science and philosophy and art of transformation seemed hopelessly entangled. In different ways all asked similar questions: What is an individual? Why must we struggle to change? Does the life cycle take us forward, or backward, or are we just standing in place?

Metamorphosis: A Natural and Human History attempts to address these questions. At its heart are two mysteries. First, what is metamorphosis, how does it work, and why does it exist? These are scientific questions whose solution we have been slow to discover. Already the ancients recognized the puzzle. And gradually, from mayflies flitting above lazy spring rivers to anchored sea squirts in frigid ocean depths, the life cycles of countless creatures were described over the centuries. In kissing bugs and immortal jellyfish, the last one hundred years have uncovered hormones controlling metamorphosis, and in the past decades, the genes behind them have been discovered and their pathways mapped. The experiments have been dazzling, matched only by the colorful scientists behind them—from an Englishman who decapitates insects, to a Japanese man who sings karaoke to baby jellyfish, to Sigmund Freud searching in vain for eel testicles, and all the way to the ancients, searching for the origins of life in cheese. Still, metamorphosis remains elusive. There are parts of the story that may forever remain unknown.

The second mystery of this book is about the meaning of metamorphosis for us humans, the questions we perennially ask ourselves about transformation and change. As a cultural preoccupation metamorphosis has attracted the attention of wordsmiths from Orwell to Ovid, philosophers from Plato to Parfit, and followers of every god or son of god from Jupiter to Jesus to Jagannath. In modern times it's spawned plays like *Peter Pan*, inspired music by Strauss and paintings by Dalí and children's books like *Pinocchio*, to name but a few. Alongside being a gateway into biology, metamorphosis in a more lyrical sense is a portal into our innermost obsessions. Why we yearn to grow but are also afraid of it. What it feels like to be ourselves. How it's possible that we remain the same while changing all the while.

We do science for practical reasons, like better medicines and weather forecasts and more efficient agriculture, and electronics. Often we do it to find "truth." But we also engage in science to encounter metaphors that help explain life to us. And as creatures that grow, adapt, and remember, metamorphosis is a puzzle that speaks to us in a very personal way. Sometimes, when we look at ourselves closely, we disband, like a caterpillar in a chrysalis. Just as often, when we look to our past, we say: "I was a different person then." But on most days we feel like ourselves, despite the fact that we're constantly changing. And so it means something to us to learn that grown butterflies have memories from the time they were caterpillars. It means something that starfish can exist as two separate, different selves. What identity is will probably forever remain a secret to us. But each of us will spend a lifetime trying to figure it out.

Metamorphosis: A Natural and Human History is divided into three parts. Part 1 hearkens back to thinkers who began to unravel the mystery of reproduction and growth. The story begins with Aristotle and moves through dramatic breakthroughs in the second half of the seventeenth century, including the discovery of the very ingredients that give rise to life. Many men appear in the tale, but its central figure is the artist and naturalist Maria Sibylla Merian, sometimes referred to as the first ecologist. In a world struggling to come to terms with the place of Man—and Woman—in God's universe,

Merian's life is a performance of the period's emblematic question: *Where do we come from?*

Part 2 picks up the story in the nineteenth century and the era of Darwin. With the birth of evolutionary theory, it now became possible to interrogate life in a completely new way. No longer perfect creatures, finished by the hand of God, species were now seen to be in a state of constant flux. And as scientists unraveled their deep histories, they also began to imagine a better future. Turning evolution into a theory of everything, they applied it to sex and race and morality. One such scientist was a brash and brilliant man named Ernst Haeckel, a zoologist, marine biologist, philosopher, and artist. Trying to overcome a great tragedy, Haeckel created a new religion, and came to believe that, because he knew where we came from, he could answer the question *Where are we are going?*

Today we can manipulate genes to control metamorphosis, and new imaging technologies allow us to peer into a chrysalis and see how a butterfly is constructed from goo. Emboldened by these powers, many believe the views of ancient Greeks, early-modern Christians, and nineteenth-century romantics can be safely consigned to the past. But anchoring Part 3 is an extraordinary husband-and-wife couple, who are using the tools of modern biology to hearken back to an ancient belief. Lynn Riddiford and James (Jim) Truman have devoted their lives to uncovering the molecular pathways of metamorphosis, showing how genes and hormones and nerves shape bodies and behaviors. They add mesmerizing detail to our biological understanding, but also expose its limits. Can science provide answers to our most intimate questions? Riddiford and Truman provide the theme of this book in its final trimester: *What is the self?*

In the dramatic episodes that follow, the story goes back and forth between conventional and imaginative biography, natural creatures and intellectual history, science and philosophy and art. A spirit of loose-jointedness blows through the book to reflect the way I experienced writing it, circling around my chosen subjects and the questions they studied with the uncertainty of a fellow curious heart. I tried to sketch the debates they inherited, and to make myself live within those intellectual and spiritual confines. I circled around my protagonists again, and again, trying to get inside their heads, even adopting their language. Along the way, in a spirit of camaraderie, I recorded my own moments of wonderment.

As I awaited becoming a father for the third time, the questions posed by metamorphosis enveloped me: Where do we come from? Where are we going? What is the self? With the help of my companions, I saw how these questions were echoed in the history of science. For me this particular journey ends with Sol, our third baby. But I'd be delighted if readers felt inspired to continue the journey for themselves.

PART ONE

WHERE DO WE COME FROM?

1.

SURINAM

It was so hot, she thought she might faint. The previous night, a hole in the netting stretched around her bed had allowed mosquitos to gain access, and they'd kept her up until dawn. She was tired, and covered in bites. But she was preparing for a deeper journey into the jungle.

She'd already learned: Traditional methods didn't work here, where she could not know whether to look down to avoid tripping on furious tree roots, or look up to avoid a giant boa constrictor falling from the forest canopy onto her head. For over thirty years she'd perfected the method: Find a caterpillar, note and collect its food plant, put it in a box, make sure it was comfortable and feed it, then wait to see what happens. Keep on doing it, time after time, until you discover the pattern of things: which egg turns into which caterpillar, which feasts on which plant before spinning a cocoon or building a chrysalis.* If the pupa produced parasitic flies or wasps, draw them

* Butterflies build a *chrysalis*, which is a naked pupa and comes from the Greek word for "gold." Moths, on the other hand, spin a protective silk enclosure around their pupa, called a *cocoon*, which is from the Old French word *coque*, meaning "shell."

and their cocoons and their larvae. By meticulously following this regimen, she'd gained a name for herself, producing her first publication on insects in 1679, a quarto titled *The Wondrous Transformation and Particular Food Plants of Caterpillars*, and a sequel four years later. But now she was in a far-off land just above the equator, a land of red peppers and papayas, palm-sized tufts of fluff that "looked like dandelions," venomous forest pit vipers, fifty-meter trees necklaced with coconuts, rivers full of hungry schools of piranhas. The scale was unimaginable, the vegetation as dense as it was lush, and however much she tried to obstruct them, wood lice would not stop falling down her blouse. Running after a butterfly here, she thought, was nothing like simply holding out a finger near a flower at the botanical gardens back home in Europe. Still, she wasn't going to sail upriver for naught.

Packed with sugarcane fields on the northeast shoulder of South America, Surinam (known today as Suriname) had been fought over for years, until the English traded it to the Dutch for Manhattan in 1667. One share now belonged to the city of Amsterdam, another to the Dutch West India Company, and a third to Cornelis van Sommelsdijck, who became its first governor in 1683 only to be brutally murdered by his own mutinous garrison five years later. Sir Walter Raleigh had traveled to this "indescribably bewtiful" tropical paradise, as he called it, losing a son there, and following him, so did the politician George Warren. The landscape, the latter wrote, was "high, and mountainous, having plain Fields of a vast Extent, here and there beautified with small Groves, like Islands in a Green Sea; amongst whose still flourishing Trees, 'tis incomparably pleasant to consider the delightful Handy-works of Nature." There were silver-beaked tanagers in this land, and anteaters and crocodiles, even the much-coveted exotic European rarity, the pineapple. Warren invited brave souls to seize the economic opportunities, along with the scantily clad Amerindian girls. The most daring, like Raleigh, had hoped to find gold in the mythical city of El Dorado. But Maria Sibylla Merian, fifty-two years old, had other plans. She had come to investigate the butterflies.

Maria and her daughter Dorothea arrived in the capital, Paramaribo, in August 1699, settling into one of the five-hundred-odd wooden houses, one with a small garden. This they used to grow plants collected in the forest just beyond the settlement, with its protective fort and cannon. Some they recognized, like the costus, with its bright-red, pinecone-like, fleshy leaves and delicate, tapering, white flowers, others remained unknown; not even the locals, Maria noted, were "able to tell me its name or properties." There were few amenities in the house, still fewer physical comforts. But as the weeks and months passed, life settled into a tropical rhythm, the sun bouncing off the crushed seashell–paved streets, the patter of rain punctuating early evenings and dawns.

Now, in the spring of 1700, she and Dorothea were about to set out by river ferry, then canoe, to the farthest outpost in the jungle. The tiny colony of La Providence had been built by a community of Labadists, followers of the French-born Protestant religious leader and pietist Jean de Labadie. As it happens, Labadie's wife had been the sister of the murdered governor, Cornelis van Sommelsdijck. When de Labadie died in 1674, his followers, including three of the governor's daughters, set up an egalitarian monastic community in the van Sommelsdijcks' stately home—Waltha Castle—in Wiuwert, Friesland. Maria and her mother and daughters had been part of this community. Although it didn't last, two of the governor's daughters decided to establish another Labadist settlement, this time in the New World. Paddling into the farthest reaches of the Surinam rainforest, they hoped to find God in La Providence.

Maria knew Surinam was as cruel as it was beautiful. Over 90 percent of its inhabitants were slaves, and the colonists treated them like dirt. The Amerindian women told her how they'd drink peacock flower seeds to induce abortions after having been raped by white men. In the book that would eventually come, *The Metamorphosis of the Insects of Surinam*, she illustrated the peacock flower and wrote beside it: "Indeed they even kill themselves on account of the usual harsh treatment meted out to them; for they consider that they will be born again with their friends in a free state in their own country, so they told me themselves." But even Maria could not escape the structural injustice. In an inhumane system, the best she could do was make a modest plea for a modicum of humanity.

Passing leopard-spotted stingrays buried in the muddy river bottom and spectacled caimans patrolling the banks, Maria and Dorothea's ferry reached Cassipora Creek, where slave-owning Jews, themselves refugees as well as colonists, had built Bracha v' Shalom (Blessing and Peace), one of the earliest synagogues in the Americas. Next came the Hernandez Plantation, then the Castilho Plantation, then there suddenly it was, La Providence. Maria had been forewarned: Some years earlier a large group of slaves had broken away from here, escaping their brutal treatment by the pietists. Women and children scrambled across the river, heading south, away from the ocean, joining other "maroons" in the jungle depths. There was always a danger the settlement might be attacked, to pillage guns or rescue relatives. Even in Surinam, La Providence had a vicious name.

The reunion with the remaining Labadists—"the satellite of a vanished planet, orbiting around emptiness," as Merian's modern biographer Kim Todd put it—must have been awkward. "Did they find the flame of belief hard to keep lit here in the understory," she asks, "where tests would come in the form of deadly disease rather than a coat with shiny buttons? How many times did they have to shield their Bibles from wood ants? So many of the others had given up and gone back home." And what did Maria make of this, among members of a group to which she once belonged, sitting down for supper meals undoubtedly punctuated by awkward silence? Having left behind her, besides her published books and work journals, only eighteen rather unrevealing letters, we will never know.

Still, Maria had business to attend to. Already she'd discovered that the wings of the Menelaus blue morpho butterfly—which she described as "polished silver overlaid with the loveliest ultramarine, green and purple. . . . Its beauty cannot be rendered with the paint-brush"—looked like roof tiles under a magnifying glass. That vine sphinx moth caterpillars ate "voraciously," and that the legs of the flag-footed bug fell off when she touched them, no matter how carefully. She was observing so closely that she was already able to refute a pronouncement of the great microscope man from Delft, Antonie van Leeuwenhoek, on the existence of eyes on the bodies of caterpillars (they didn't really exist). But her appetite could not be satisfied, and she pushed ever farther into the jungle.

"The forest grew together so closely with thistles and thorns," she wrote of her excursions beyond La Providence, "I sent slaves with hatches ahead, so that they chopped an opening for me, in order to go through to some extent." Here, virtually all the plants were unknown to her, and most of the insects were complete novelties. There were ants that made bridges with their bodies, she wrote, and others that "can eat whole trees bare as a broom handle in a single night." Hairy tarantulas sucked blood out of hummingbirds. Giant water bugs devoured frogs, and snakes burrowed through cassavas to lay their eggs inside. There were strangler vines, and buttress roots, and cactus sprouts, and tiny worms and spiders and geckos, all entangled. Doing her best under the circumstances, she sketched each of these in her notebook, always life-size, always depicted within its habitat. Later, safely back home, she'd carefully etch a copperplate engraving, run it through the press to create a partial print, and when that print was still wet, transfer a reverse image unto vellum. Then she'd unfurl her wooden brushes, take out her gum arabic, and begin painting in watercolor by hand.

In La Providence, Maria recorded an enigma that had puzzled her for months. Back in Paramaribo, she and her lady servants were woken up in the middle of the night: A loud chirp echoed in the house. Following the noise to its origin, they found a box, and upon opening it, as Maria records, "a fiery flame came out." They jumped back in fright. It was the bulging, alligator-like heads of the lanternflies in the box that had lit up, she thought, a sight like no other. But where had the lanternflies come from? It was almost impossible to catch a young lanternfly to chart its metamorphosis; they were silent, cryptic, nocturnal, and rare. But Maria had captured a little immature green beetle, a cicada, whose wings seemed to grow slightly larger with each molt, and who made a sound "like a lyre." The local women told her that this insect was the "mother" of the lanternfly, but could Maria be sure? In bed, in the hinterland of La Providence, she heard the distant chirp in the darkness, and wondered.

2.

CICADA

There was no way Maria Sibylla Merian could have guessed, but cicadas have nothing to do with lanternflies, and have been chirping away before the dinosaurs arrived, besides. They have continued to chirp through five mass extinctions and five ice ages. Appropriately enough, they happen to be one of the planet's slowest maturing creatures.

In the famous mariachi song "La Cigarra" ("The Cicada") by Raymundo Pérez y Soto, cicadas are romanticized as creatures that sing until their death. Another Latin American tune, recorded by Mercedes Sosa, "Como La Cigarra," hails the cicada as a symbol of defiance, inspiration for eternal survival. The Greeks seemed to have grasped this double nature, already in the times of Homer: In the *Hymn to Aphrodite*, Eos, the goddess of the dawn, begs Zeus to make her lover immortal. Zeus grants her wish, but she forgets to ask that he also become ageless. The lover grows old and never dies, shriveling and shrinking, until he turns into the world's first cicada.

Still, the ancient Greeks called them the "love of the muses," the "sweet prophet of summer," and likened them to gods. Athenian ladies wore gold cicadas in their hair as ornaments, and they were kept in cages, like pet birds. A toy cicada sitting on a harp was an emblem of the science of music, and while the acerbic Homer compared them to garrulous old men, Antipater preferred the notes of the cicada to the swan's:

> *The Muses love thy shrilly tone*
> *Apollo calls thee all his own*
> *'Twas he who gave that voice to thee*
> *'Tis he who tunes thy minstrelsy.**

Cicadas accompanied humans through their joys and sorrows, and many cultures developed unique traditions involving them. In Suriname, as reported two centuries after Merian in the *American Naturalist* by the medical doctor F. C. Clark, young boys fasten straws to cicadas and run with them through the streets.

Where do cicadas fit in nature's economy of change? A minority of insects, like silverfish and bristletails (and also earthworms and lice), undergo no metamorphosis at all. They are called *ametabolous*, and once they hatch from their eggs they sport all their necessary body parts and simply grow larger with every molt. Cicadas, on the other hand, experience something more dramatic. At their final molt, to finally become true adults, they sprout sex organs and true wings that will allow them to take to the skies and find a mate.

There are two major cicada clans: a small offshoot in Australia and Tasmania—including the Australian greengrocer, one of the loudest insects in the world—and all the rest, three thousand species strong, strewn across the planet. Only Antarctica is innocent of this seemingly carefree insect. Today we know that there are fully fifteen species from nine genera, three tribes, and two subfamilies of cicadas living in Suriname; the one Maria ended up painting is known by its Linnean name, *Fidicina mannifera*. A cicada is a rather unusual-looking creature. Its compound eyes, placed far apart from each other on either side of two small antennae, give it a strangely

---

* Translated by the nineteenth-century Irish poet Thomas Moore.

earnest gaze, like an extraterrestrial's, or—if you look at it long enough—a vacuous one, like a cow's.

Cicadas come in all sizes. Some species, like the empress cicada, are huge, with a wingspan of eight inches, others are smaller than a dime. And humans have put them to many uses: to forecast the weather, as medicine, or money; as delicious meals, especially as grubs, deep fried in Shandong cuisine. Their many species are colloquially known by names like black prince, double drummer, and whiskey drinker, which makes sense when you recall that in Hesiod's *Shield* they are mentioned as those who sing when the millet ripens. Chirping in a chorus of males up to 100 dB high, a cacophony equivalent to a power lawn mower or garbage truck, cicadas can bring about permanent hearing loss if you get too close for too long.

The surface of their wings are a fortress. Covered by tiny, waxy spikes, they repel water and rip open the membranes of bacteria, making cicada wing surfaces the world's first-discovered bacteria-killing biomaterial. No less incredible are the rubbery structures that allow for singing: a pair of tymbals straddle the abdomen on each side like a saddle. Rapid alternate flexing and relaxation of the stomach muscles buckles and unbuckles these bells, producing the cicada's music. Males of some species have almost entirely hollowed out abdomens, which act as sound boxes; others use their tracheae as resonance chambers to amplify their song. Some rub their wings over a series of ridges on their thorax to add a further chorus line. Singing is everything in the world of cicadas. And each species has its own love melody to bring on mating.

What are called annual cicadas appear each season, usually in the late spring or early summer, when temperatures in the shallow soil reach about 64 degrees Fahrenheit. It's then that their nymphs emerge, having lived for some time at depths down to eight feet in the ground. In annual cicadas, the life cycles of these nymphs vary from two to ten years, but they don't all mature at once, so only a fraction appear each year. Periodical cicadas are a different story. These species—of which there are seven in North America—gradually develop to full sexual maturity over five molts, reaching the finish line together every thirteen or seventeen years, depending on the species. Then, like an army of synchronized swimmers gone mad, they break through the soil all at once, as many as a million and a half individuals per acre, all of them looking for a branch or trunk on which to molt one final time.

During the previous molts the tiny threads securing the creature's skin to its outermost layer, or *cuticle*, detached, and the gap was filled with a gel. Increased cell division created folds in the surface of the skin that allowed it to expand like an accordion. The skin built a new cuticle while the gel absorbed the old one, and as air filled the gap, the insect stepped out of its exoskeleton, a perfect copy of its former self, only larger. A waxy layer then coated its expanding skin all over again, and the new cuticle hardened, like an armor. This was a ritual experienced five times underground, in total darkness.

If you've ever neared a tree that hosted cicadas, you'll have seen the empty husks, called *exuviae,* littering the trunk, clinging eerily to the bark, like ghosts. Until you really get close to them, you think they're alive, which explains why the Chinese use the phrase "to shed the golden cicada" when they speak of fooling one's enemy by using a decoy. Clawing out of them, allowing their new form to harden and darken from its teneral shade of white, the cicadas assume the final stage of their life cycle. In the next three weeks, males will sing, females will click, and sex will be had. Females will proceed to cut slits in the bark of twigs and carefully deposit hundreds of fertilized eggs. And all will then die in great numbers.

The torch has been passed to the next generation. Within four to six weeks, the eggs will hatch, and baby nymphs will drop to the ground, where they'll burrow into the soil. Feeding on the xylem sap of the roots of the trees where they were conceived, they'll begin the cycle of slow molts and growth underground. After thirteen or seventeen years, when the temperature of the soil reaches exactly 17 degrees Celsius, they'll perform their synchronized breakout. Then they'll sing, have sex, and expire, and the cycle will start again.

Why, as the medical doctor Clark wrote in 1875, does the seventeen-year species "take his leave, and testifies with his lingering life, a glad song which grows feebler and feebler, till finally it dies away sadly but beautiful like the summer he carries with him?" Why the excruciatingly slow maturation, the frenzy of singing and mating, followed by swift death? Both Benjamin Franklin and Thomas Jefferson pondered the question, the latter remarking that the appearance every seventeen years reminded him of great locust years at Monticello. The Swedish taxonomist Carl Linnaeus gave the apt name *Magicicada septendecim* to a specimen he'd been gifted. But why seventeen? How weird. No one knew the answer.

Today, different theories abound. Periodical cicadas emerge every thirteen or seventeen years, one goes, since no predator exists who could possibly depend on such a diet. Another, to the contrary, speaks of "predator satiation survival strategy": If massive volumes of cicada emerge all at once, those creatures who feed on them will have their fill, ensuring that the rest of the cicadas survive. Yet a third explanation harnesses mathematics, and the wiliness of prime numbers: emerging every thirteen or seventeen years was an adaptation devised in between ice ages in the Pleistocene, to prevent broods of cicada with different cycles from hybridizing when natural selection was working on them very strongly. Later, the magic of prime numbers would be co-opted to prevent potential predators from receiving periodic cicada protein boosts by synchronizing their own generation cycles to devisors of the cicada's time of emergence.

Maria Sibylla Merian knew none of this, and, inevitably, she sometimes made mistakes. One of the gravest of these had to do with the cicada. Unable to work out the life cycle of lanternflies, when she finally painted them she included the "mother" the locals had mentioned to her, in between the lanternfly and the cicada. There was no way she could have witnessed this, because cicadas are unrelated to lanternflies and don't give rise to them. To make matters worse, modern biologists now know that lanternflies don't emit light from their heads at all. So what was the "fiery flame" that had scared Maria and her women servants in the middle of the night? Kim Todd suggests that Maria glimpsed fireflies trapped in the box, or dead lanternflies whose heads harbored bioluminescent fungi. Her mistake may be

Maria Sibylla Merian, Branch of pomegranate with lanternfly and cicada, 1702–3, Royal Collection Trust.

forgiven: Young lanternflies bear a resemblance to cicadas. In fact, the spotted lanternfly is still referred to in the East as "chu-ki," or the spot clothing wax cicada. It would take two hundred years for naturalists to dispel the myth.

Imagining all this, I start to wonder: Who was this fifty-two-year-old, seventeenth-century white woman, walking behind Black slaves hacking through the jungle with sketchbooks and butterfly nets and insect boxes and hidden dreams? Where had she come from? Her arrival must have seemed as mysterious to the locals as the appearance of millions of cicadas every season is to us. In fact, to understand Merian and her cicada, we need to go back thousands of years—to the ancient mystery of where life comes from.

And to the world's first biologist, Aristotle.

3.

ARISTOTLE

I remember the giggles in class: My eighth grade teacher told us that Aristotle thought women had fewer teeth than men, apparently not having considered asking his wife to open her mouth. Actually, the man whom Bertrand Russell called the first to write like a professor got a lot more wrong from our perspective. He confused the functions of heart and brain, claimed that heavier objects fall faster to earth than lighter ones, and imbued cucumbers with "souls." After centuries during which Aristotle dominated Western thought, the Scientific Revolution and its heroes came to reject many of his teachings. Yet we owe more to Aristotle than we can imagine. Perhaps his greatest legacy was the choice to look closely at the things of this world rather than searching for truth, like his teacher Plato, in the realm of pure ideas. Taxonomy was born from this empirical disposition, and with it the beginnings of a scientific method. It all began in a resplendent lagoon on the island of Lesbos, where Aristotle opened his eyes to examine (mainly) sea creatures. We have the cuttlefish and the periwinkle to thank for the fortuitous descent from the world of Forms.

But Aristotle also looked at insects. And more than any other creatures, they threatened to drive him mad. After all, had anyone ever seen where flies or bees came from? They seemed to appear from thin air. In his book *Generation of Animals*, Aristotle claimed that wherever there are two sexes, it's the male who provides the shape of the offspring; the female, on the other hand, just provides the matter. In humans this meant that sperm carried with it everything that really mattered—the very essence of a being, its "form and knowledge," or "soul." The menstrual blood, on the other hand, was just there to be molded: Shortly after it combined with the male ejaculate, Aristotle thought, following Hippocrates, a heart was born in the womb, signaling new life.

How did sperm do all this? Aristotle claimed it brought with it a *Pneuma*, or breath. This was a kind of heat, or energy—Aristotle called it *dynamis*, which literally means "capability" or "power"—as opposed to something material. Just as a carpenter gives shape to a piece of wood without himself adding any matter, so male semen brought order to the chaotic matter provided by the "female semen," whose main attribute was that it had the potential to be sculpted, a process Aristotle called "fetation." Today we know there's no such thing as "female semen," and we distinguish between the various parts of the process—conception, reproduction, and development. But "generation" was the single term Aristotle used for all three. Once the semen met, the rest unfolded. In some places he likened it to cooking, or the way fig juice was known to curdle milk. That was the explanation for why sex was necessary. It was the act that sparked the process that gave shape to life.

Except that there was a problem. Some creatures, including eels but also most insects, were never seen having sex. Supposing they never copulated, how could these creatures receive a *Pneuma* from a father? How could they ever take form? Aristotle's answer was: metamorphosis.

It was all about perfection. A creature that gave rise to something that looked and behaved like itself was "perfect," or "finished." Besides growing bigger, the new creature had no further reason to change. By contrast, a creature that was generated without sex always had an essence, or *Eidos*, different from its parent's. It would need to change from the incomplete or "imperfect" form of the larva* to the complete or "perfect" form of the adult. Before such a creature could become like its parent, it had to gain a new essence, to literally transform.

And so Aristotle created a hierarchy, based on whether and how creatures bred true. He placed humans at the top above mammals—both produced live babies just like themselves, he thought, without the need for an egg. Below them on the scale of perfection came creatures that produce live young with eggs retained in their bodies, like some fish and frogs. Then came those who produce "perfect" eggs outside their bodies, which stay the same size until they hatch, like reptiles and birds. Below them were cephalopods and crustaceans, copulating creatures whose "imperfect" eggs actually grow outside the body. And finally, at the very bottom, came insects, or at least most of them, which produced wormy larvae without any trace of ever having had sex. Such larvae, Aristotle wrote, were actually "soft eggs," and "imperfect"; if they survived they would harden into a pupa,** which was like a "perfect" egg of a hen. And if that survived, having finally found its form through the direct workings of air and heat and a suspiciously semen-like "foamy" moisture, the creature could turn into an adult.

Aristotle wanted it made clear: There was an analogy between sexual and spontaneous generation, but never quite an identity. And though he wasn't at all known for his modesty, the mysterious way in which formless material within an egg turns into an animal seemed to him "no small puzzle." Just as in plants, insects presented countless

---

* The *larva* gets its name from the Latin for "mask," implying that a genuine nature lingers hidden within. Whether that's true or not, the ancient etymology also links to "ghost," or "hobgoblin," hinting at ephemerality, intangibility, or perhaps simply mischief.

** A *pupa* carried its Latin meaning into the German for "girl," or "doll." In English it became a "puppet," waiting to be animated, indicating a liminal stage between immature and mature form.

exceptions. But in the future tug-of-war between the doctrine of *preformation*, which held that a preexisting template imposed form on the materials of the developing embryo, and the doctrine of *epigenesis*, which held that the materials somehow self-assembled, Aristotle held tight to both ends of the stick: The creation of form was a meeting of matter with action, and took place gradually from within. But always there needed to be an external breath, the *Pneuma*. Life couldn't come into being without an initial spark.

This was all well and good, except that Aristotle had noticed that some insects actually behaved more like birds: Rather than first producing "soft eggs" in the form of caterpillars or maggots, they produced "perfect eggs" that hatched smaller versions of themselves. These were the copulating insects, with the same name and nature as themselves, and they included locusts, spiders, wasps, ants, and... cicadas. Never mind that Aristotle was wrong about wasps and ants; locusts, spiders, and cicadas (and also crickets, cockroaches, and earwigs) do, in fact, belong to a group of insects that look like smaller versions of themselves when they're born and grow from molt to molt until they finally sprout functioning wings and sex organs. Aristotle was certain that these insects don't metamorphose, since they were "complete" already, with the same essence, or *Eidos*, as their parents. We now call them *hemimetabolous* insects and refer to their transformation as "incomplete metamorphosis," a bit confusingly. The *holometabolous* insects whose transformation we refer to as "complete metamorphosis," on the other hand—like wasps, ants, silkworms, bees, beetles, and butterflies—are Aristotle's "imperfect" insects, the ones appearing spontaneously, without parents, or sex.

But wait. Aristotle actually saw butterflies mate. He saw gnats and mosquitoes mate, and even snails mate, too. We know this because he told us. Yet he insisted on claiming that they were "imperfect" for having no mothers and fathers. What looked like mating was just playing, he argued, what looked like fertilized eggs were nothing of the kind. As in the case of women's teeth, Aristotle seems to have been blinded by his own theories. How did the great philosopher commit such a blunder?

One answer is that he had been trained to mistrust his senses. The idea that appearances could deceive was stamped into Aristotle's culture. Ancient Greeks would tell their children the story of how the Titan Prometheus made humans from clay, but unlike the story most of us know, theirs had a twist. When Prometheus decided to sculpt the form of Truth (*Aletheia* in Greek) in order to keep humans honest, he was called away by the mighty Zeus before he could finish and left his studio in the hands of an apprentice. The apprentice, Dolos (Trickery), fashioned a figure of the same size, but ran out of clay before he could make its feet. When Prometheus returned, he gawked unbelieving: His and Dolos's statue were practically identical! Unable to tell the difference between the two, he put them into the kiln and infused both with the breath of life. It then became clear to him that while Truth walked with measured steps, her unfinished twin stood stuck in her tracks. The forgery acquired the name Pseudologos (Falsehood), a reminder that while lies might start off successfully, in the end Truth overtakes them. By doubting the evidence before his eyes, in favor of an elegant theory, Aristotle was doing the intellectually prudent thing.

A second answer as to why Aristotle might have blundered is that he never thought to put his ideas to the test. He certainly observed carefully and tried to proceed in a systematic way, but the idea of a controlled experiment remained foreign to him. It would take centuries for the experimental method to begin to distinguish science from other kinds of wisdom. And while we might have many ways to work out whether we can or cannot trust our senses, it seems unfair to expect Aristotle to transcend time itself.

Furthermore, the doctrine of spontaneous generation had been accepted for thousands of years and for understandable reasons. Lacking a concept for either what a species was or what constituted inheritance, it made sense for natural philosophers to believe that life could emerge from just about anywhere: worms from cheese, moths from wool, toads from ducks putrefying on a dung heap. Before Aristotle, the sixth-century BC Greek philosopher Anaximander had claimed that warm water and earth gave rise to fish, inside of which human embryos were held prisoners until puberty, upon which they'd burst out as men and women, capable of feeding themselves. Anaximenes, after him, held that all plants and animals, including humans, came from slime. And a few centuries after Aristotle, the Roman architect Vitruvius counseled that libraries should always be built to face east; otherwise, westerly and southerly bookworm-generating winds would destroy their treasures. Not until well into modern times was there any compelling reason to believe that like must necessarily give rise to like. A woman's damp hair could spontaneously turn into snakes if the weather got warm enough. And rotting tuna could give rise to maggots, some believed, which turned into flies and then grasshoppers, before finally taking the shape of quail.

The ancient world was suffused with miraculous transformations, and thinkers gradually turned them into a science. The Roman poet Virgil counseled honey makers to bury a bull with its horns aboveground, wait for nine to thirty-two days, and cut the horns: Supposedly, bees would come flying out. Pliny the Elder wrote in his *Natural History* that silk moths came from the vapor on rain-fallen oak blossoms, and that sun-washed dewdrops on radish plants shrank to the size of seeds, giving rise to maggots. Like many who came after him, he would have read the works of his contemporary Ovid about dragon teeth sown in the ground turning into a ferocious army of warriors, or a huntress seduced by a god, then turned into a bear, then a constellation. Or the boy consumed by self-love, staring into a pool at his reflection, a myth that would fascinate the twentieth-century surrealist painter Salvador Dalí. Alas, the boy would not be able to part from himself, turning into a flower.

In retelling the myths of the Greeks and Romans, Ovid followed tradition. He, too, was taught the wisdom of the ancient philosophers. "All is in flux," he writes,

> Nothing retains its original form, but Nature, the goddess of all renewal, keeps altering one shape into another. Nothing at all in the world can perish, you have to believe me; things merely vary and change their appearance. What we call birth

> is merely becoming a different entity; what we call death is ceasing to be the same.

And so, citing facts "fully established," Ovid spoke of frogs coming from mud and hornets from warhorses. "If you remove the branching claw of a crab . . . ," he warned, "and bury the rest in the earth, a scorpion will shortly emerge . . . and threaten your feet with its hooked tail."

Ovid and Aristotle and the rest of the Greeks and Romans were forgotten for a time during what used to be called the Dark Ages, but the doctrine of spontaneous generation persisted. Fishermen and shepherds, beekeepers and cheese makers, and physicians and quacks all carried on local traditions about the origins of things—and thinkers who wrote their thoughts down took pains to relay their knowledge. Augustine of Hippo, for example, a Berber who would one day become a saint, taught that sex was sinful but as unavoidable as spontaneous generation. His ideas insinuated themselves into the great seats of learning in Baghdad. And with the help of translations from Arabic into Latin, book by book, treatise by treatise, slowly the wisdom of the ancient world was recovered.

It was being elaborated on, too, expanded upon and Christianized. A German Dominican friar named Albertus Magnus systematized Aristotle's theory of spontaneous generation by affixing it to a new cosmology. In his rendering, a plethora of creatures directly acted upon by the sun, the stars, and the planets came into being from frothy foam. Through the teachings of Albertus, and the Latin reception of the works of an Andalusian polymath known as Averroes, astrology grew in importance not only as a means of prognostication but as a tool to explain the origin of life itself. More than anyone, it was a student of Albertus, the Italian Thomas Aquinas, who brought Aristotle to the Christian world. "Perfect" and "imperfect" creatures could now be made sense of through Christian doctrine. Faith and science were beginning to merge.

All the while, ideas about inheritance remained creative. Gerald of Wales, a priest who traveled to Ireland in the twelfth century, made

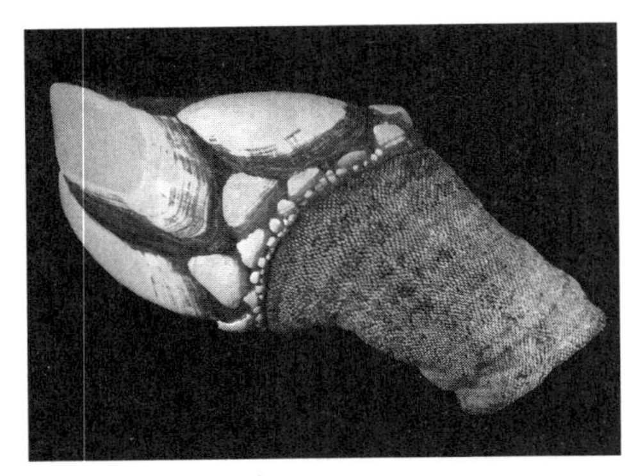

The goose barnacle.

The barnacle goose.

use of the local wives' tale that the barnacle goose came from goose barnacles to argue for the virgin birth of Jesus. This was no theoretical matter, since fasting during Lent allowed fish eating while prohibiting fowl. Gerald of Wales thought that the barnacle goose's form of reproduction provided a loophole to this culinary interdiction. Alas, his stroke of genius was batted down some years later by the third Pope to be called "Innocent." Despite his name, the Pope was not convinced by the portly, poultry-loving priest.

As the age of Petrarch gave way to the age of Shakespeare and then Rembrandt, ideas about where life came from continued to evolve. People believed that the Earth once had the power to create all life-forms spontaneously, including humans, but then cooled down. Now, in its cooler state, it retained only the ability to make creatures with cold blood, or with no blood at all, like serpents, insects, and frogs. With the rediscovery of the Greek philosopher Democritus, the seventeenth century moved away from theories of generation based on heat to ones based on atoms. Even humans could be brought to life spontaneously, modern atomists believed, by the movement of these tiny particles. It may have helped that new editions of Ovid's *Metamorphoses* were now being widely circulated thanks to Gutenberg's printing press. Inspired by his tales, the imagination of the moderns began to surpass even that of the ancients.

And so, well before Aristotle and well after him, people thought that life could come from nonlife, and many other places. Considering the great philosopher and his "blunder," perhaps it's best to recall that for a thousand years the smartest people looking into the skies were sure there had to be epicycles to prescribe the movements of the planets. That there had to be a celestial ether. Scholars have taught us that these imaginaries were necessary to salvage the Christian dogma that Earth is at the center of the universe. When it comes to what we call truth, context always matters. And no one, not even Aristotle, lives outside their times.

It was natural for us eighth-graders to laugh at Aristotle for miscounting women's teeth. Having grown up a little, I appreciate that the world Aristotle inhabited had different mores about what a woman and a man are all about; mores so far from those we hold today in the West that they constrained not only what he could think, but even what he could observe. Culture is so powerful a force it can deceive the senses.

But the greatest deceiver of all is Mother Nature herself.

4.

IMPOSTERS

Every summer you'll see him, usually in July, fluttering about in an open field. From Spain to Armenia through Kazakhstan to Siberia, and all the way to Sichuan, the large blue butterfly (*Phengaris arion*) is aloft. Like symbols of nature's beneficence, or living examples of art, they seem to flutter everywhere without a care in the world.

I didn't know this as a child, watching them hatch in my dry aquarium, but the large blue butterfly was, and is, in serious trouble. First recorded in Britain in 1795, it has since gone extinct there, as well as in Belgium and Holland, and populations are dwindling in Denmark, Germany, and France. The UK Biodiversity Action Plan has succeeded in bringing back small numbers and so has a project in the Ordesa Valley in Aragon sponsored by UNESCO. While scientists remain unsure about why this is happening, across Europe the large blue is fighting for its life.

To stay alive the large blue uses, and has always used, a weapon, and that weapon is deception. You wouldn't guess it to watch one flitting above a meadow of wild thyme like a flying jewel, two inches across. Look closely at one of those thyme bushes. You will see that a mother blue has laid twenty to thirty flat and semiglobular, pale blue eggs in the purple blossoms of the thyme. When the eggs hatch, tiny ocherous larva caterpillars emerge, with a lilac tinge on their sides. Until autumn they will nibble on the thyme, and like the cicada undergo several molts. At the fourth molt, they drop to the ground. Then the tiny caterpillar begins to secrete a pheromone. Before long, groups of little red ants of the species *Myrmica sabuleti* hesitantly come by to sniff it, picking it up and lugging it to their nest. There, in the nursery chambers, they will loyally tend to the tiny caterpillar, feeding it and cleaning it as it grows to more than one hundred times its original mass.

The large blue caterpillar has tricked the ants into believing it is one of them. Not only by the scent it makes, which perfectly mimics the scent of their own larvae, but also by means of acoustic chicanery, producing a scratch-like noise when begging for food. The noise sounds just like the one made by a larval queen, and the worker ants give it preferential treatment at the expense of their own kin. Even though it's much larger than they are, and a different color, the growing larva of the butterfly continues to be nurtured by the ants as it builds its chrysalis, climbing out again in June with legs and wings, ready to take to the skies.

Parasitism is a fine-tuned affair. If the mimicry is not perfect, the ants become suspicious, leading to almost certain, and brutal, death. Sometimes the cover is blown when a caterpillar is found by the wrong (though closely related) species of ant. Even small deviations from protocol will end in disaster. Research in this vein has borne out something called the "queen effect": Large blues are three times less likely to survive in a *Myrmica* nest if there is a queen present among them. The reason for this is that ant queens lay eggs the fate of which depends on the queen's condition: If she dies, or otherwise leaves the nest, worker ants will groom the largest of the larvae as they transition into "virgin" queens. But if the queen is healthy, and present, she instructs the workers to neglect the growing eggs, starve and even bite them, disrupting their growth so that they emerge as workers rather than "virgin" queens that would challenge her. The large blue imposter has to be painfully attuned to its surroundings, maintaining a strict balance between mimicking the queen in the presence of workers and appearing to be a worker to avoid the wrath of the queen.

These court intrigues were far from my mind when the large blue first captured my imagination. And, before long, so too was the large blue itself. Why people lose interest in one obsession and move on to the other is often a mystery, but it just so happens that for me butterflies were replaced by Russia and Catherine the Great. For a time I read everything I could about her, dreamed about her, spoke to her, imagined myself walking, even sleeping, unnoticed by her side. Born Sophie von Anhalt-Zerbst in Prussia to an impoverished prince and a mother with a well-regarded bloodline, fate took a twist on her: Russia was desperate to continue its line of emperors and needed someone to produce an heir. When the daughter of Peter the Great assumed the throne in a coup in 1741, unmarried and childless, she chose her nephew Peter as the man for the job and went about searching for a bride. Peter needed an empress from the aristocracy, but not from a family of any power or political maneuvering. Sophie was the lucky choice, traveling north to Russia by horse and chariot to become his wife at fifteen. Converting to Orthodox Christianity shortly after, she became Ekaterina, or Catherine.

Catherine, it is said, had high expectations of love, all of which shattered the moment she met her liege. He was crass and devoid of gentleness, and nearly nine years passed before an heir was seeded. The courtiers gossiped that he was the son of a military officer, not Peter, and Catherine did little to deny it. Maybe this was a form of revenge served cold: She herself thought her husband, however much she hated him, was the father of her son.

When Peter the Great's daughter died in 1762, her chosen nephew became Czar Peter III, and Catherine became his consort. By now so utterly disgusted with him, and proficient in Russian, she determined to transform her childish romanticism into an adult love for Mother Russia, doing whatever it took to gain her place. Just six months after her husband had become czar, Catherine hatched a plan to dethrone him. "All his actions bordered on insanity," she wrote in her memoir, and she needed to save the country. Her conspiracy was uncovered, but she moved quickly and was smart. Within days she forced her husband to abdicate. They called it a "bloodless coup" despite the fact that on July 17 Peter was found dead. The official version was that he died of hemorrhoidal colic, which soon became a euphemism for assassination. Most probably his death came at the hands of a brother of Catherine's current lover. Ever since, Catherine's enemies would grow in number. She despised them, and they considered her an imposter.

Catherine had started out as a foreigner, but as the years passed, no one was more Russian. With a political genius and renowned lovemaker at her side, she and Prince Grigory Potemkin began consolidating the empire, squashing dozens of uprisings and sending dissidents to the other side of the Urals. Using soldiers like pawns, she attacked the Ottomans, the Lithuanians, and the Swedes. She pushed down to the Crimea. All the while she became known as Europe's most enlightened leader, though many judged her a hypocrite. According to one report, court musicians were employed during concerts to tell her when to clap, but she nevertheless wrote opera libretti for enjoyment and almost succeeded in employing Mozart at her court. She drafted her own legal code and sponsored the first-ever national system of education. In the Winter Palace we now call the Hermitage, she wrote books on women's education, children's fairy tales, and amassed what many would claim was the world's greatest collection of art.

None of this mattered to her detractors, for whom she was always no more than a pretender. To them she had tricked her way into her station, taking away what belonged to them. And, like many powerful women that came before her—from Cleopatra to Marie Antoinette, Catherine de Medici to Anne Boleyn—she was made to be a harlot. Her great rival Frederick the Great said of her with characteristic misogynist disgust, "A woman is always a woman . . . in feminine government the cunt has more influence than a firm policy guided by reason." Others called her a nymphomaniac, pointing at her many lovers. Biographers report that she did enjoy lovemaking. But to her mind she had grown the wings to deserve her lust.

It seems random that Catherine would have followed the large blue as a youthful obsession, but in retrospect perhaps there was a deeper connection. After all, to Catherine's contemporaries, as well as to the large blue's duped servant ants, each was, and is, an imposter. It's said that art is a lie that helps us see the truth, but sometimes the resemblance of art to life is just uncanny: The young and idealistic Sophie von Anhalt-Zerbst became Catherine the Great, Russia's feared and celebrated ruler. Marking this dramatic transformation are two portraits: the first made of her in her youth, by the artist Georg Christoph Grooth, and a later one drawn by Vigilius Eriksen when she was reigning empress. In the first painting the sash is ocherous red and girly. But in the second it has morphed into a confident royal blue, mimicking the life cycle of the large blue butterfly.

If art can be a likeness to nature, this is hardly a coincidence. Just like Catherine, the large blue imposter has its own enemies, and they are no less devious. An ichneumon wasp will sometimes fly into the ant nest itself, stealthy and lethal, looking for a home for her own brood. As she scuttles through the opening, she is immediately attacked, but secretes a pheromone that makes the ants assail one another, forgetting about their new invader. Quickly advancing to the nursery chambers, the wasp will then proceed to identify the defenseless pink caterpillars among the other larvae, climbing atop them and injecting them with her own eggs. As the defenders continue to fight among themselves, the wasp makes a hasty exit.

And so, when summer comes around again, the unsuspecting ants may be doubly surprised. Having carefully cleaned, fed, and protected the caterpillars and then their chrysalises all winter, not only may a large blue butterfly emerge, but if the invader's egg succeeded, then suddenly—a wasp. The shell of the chrysalis will register, unwittingly, the butterfly's own betrayal: Having tricked its caregivers, it was tricked in turn.

And, for a time at least, so was Catherine. When he became czar, her son passed an edict forbidding a woman from ever assuming the Russian throne, a slap in her face. Though his own tyrannical rule would be cut short when he was strangled with a sash by assassins, he put a stop in Russia to Catherine's flirtation with the Enlightenment. Most embarrassing of all was a final humiliation: The Empress's death came about, her enemies spread the word, when the harness of a suspended stallion snapped, crushing her *in flagrante delicto.*

The story was just slander. The mundane truth is that she suffered a stroke, just like Maria Sibylla Merian, whose papers on the butterflies of Surinam had ended up in St. Petersburg, purchased by the physician of Peter the Great. Merian's own granddaughter, in fact, would have met Catherine on more than one occasion, married as she was to Russia's greatest mathematician, the one-eyed genius, Leonhard Euler.*

I had graduated from the large blue butterfly to the monarch—the human female kind—and often wondered whether I was the same person still. Perhaps the newly formed me was an imposter, a foreign creature masquerading as myself. Whether or not my musings had any merit, obsessions seem to be a part of our normal human development. Their appearance and succession are often mysterious to us, but they can become linked if we insist on it as we look back. That insistence is a narrative we tell ourselves, the one we call the story of our lives.

---

* Leonhard Euler (1707–1783) was a Swiss mathematician who spent much of his adult life in Russia. He is famous for having made influential discoveries in infinitesimal calculus, graph theory, analytic number theory, and topology, and is considered to be one of the greatest mathematicians who ever lived.

5.

THE TRANSFIGURATION OF JESUS TO CHRIST

Some would say that she underwent her own metamorphosis, having hatched from an egg in Germany, pupated in the Labadist community of Wiuwert, then grown her wings in Amsterdam, before flying as an imago to Surinam. They would say that she learned to love the insects. But the truth is, the love was always within her.

Maria Sibylla Merian was born in 1647 to a family of Belgian Walloons in the city of book fairs and religious tolerance, and on her Calvinist father's side, to one of Frankfurt's leading publishers. Her father produced books for wealthy clients on alchemy and natural history, geography and adventure—like Theodor de Bry's *Historia Americae*, illustrating the exploits of explorers to the New

World, Sir Walter Raleigh's searches for El Dorado included. But he died when she was three, leaving her his name and the emblem of the printing press, "Pious diligence wins." Quickly, her mother married Jacob Marrel.

She would remember the smell of ink, musky with hints of honey, or lime. She would spy from behind the kitchen window the poor souls who'd come from the countryside in search of a walled city, dripping with grime. There had been a never-ending war between Protestants and Catholics, and a terrible plague. In some towns, every third person had died. Still, there was something to live for, festivals and book fairs and art and nature, which was God's art. As she grew she began to see that the family press attracted free thinkers: Most others held that the only book above suspicion was the Holy Bible.

She attended school, but as a girl was never taught Latin. It was at home where she received her true education. There she would grind pigments and mix them with the sap of the acacia, to better bind them to the page. She'd engrave illustrations on copperplates; sort quills by bird—swan, eagle, goose, crow, lark—sometimes even tend to the account books. She learned how to draw and paint and etch, copying from her stepfather and half brothers. And, she would help her mother cook.

Marrel was a specialist in floral engravings, which, thanks to the tulip craze in the Netherlands, served them well. But it was the minuscule bugs in his drawings, on petals and leaves, that caught her attention. To him they were emblems: a dragonfly signifying the eternal life of the soul; bees, industry; locusts, retribution; a snail, the virtue of patience; death—a lowly fly. But she looked more closely, imagining herself one with them. Nothing made her happier than going out to find insects to be his models. Later, her mother would lament that this "odd and dirty obsession," as she called it, had been due to her eyeing a collection of bugs while Maria was still in her womb.

Understand, what we would one day call biology had scarcely been invented, and science and magic walked hand in hand. In reports of the highest caliber, sightings of new planets lay side by side with sightings of unicorns and dogs who could bark in French.

The origin of things was miraculous: raindrops bringing frogs to the world, cheese yielding worms. To believers, the transformation of caterpillar into summer-bird* was an earthly reminder of the transfiguration of Jesus to Christ. As for Maria, she searched out and found her first silkworm caterpillar at thirteen.

Aristotle remained the authority. But even he thought that caterpillars came from dew, as did Pliny the Elder, whose encyclopedias she knew. Above all, there was Ovid, a beautiful edition of whose *Metamorphoses* her Papa had published. Slipping into bed with a candle glowing on her bedside table, she would leaf through its pages to marvel at Apollo chasing after Daphne until she could do nothing to save her honor but become a laurel tree; at Zeus turning into a swan to win the love of Leda; at the nymph Io banished by Hera as a heifer across the Ionian sea. She read, too, of the crown given by Dionysus to Princess Ariadne at her wedding, tossed into the heavens and turned into a constellation. *In nova fert animus mutatas dicere formas / corpora*, Ovid wrote there—"I intend to speak of forms changed into new entities." Everything is always transforming, from the night sky to the gnat.

She adored Ovid but listened carefully to new voices. Most impressed was she by the man from Middelburg who described countless life cycles of summer-birds. Johannes Goedaert wrote for gardeners and embroiderers, natural historians and lovers of art. Carefully, in his book *Metamorphosis Naturalis*, he showed that each kind of caterpillar made a particular kind of pupa, which, when it cracked open, revealed a peculiar adult. But the caterpillars themselves, he wrote, were "bred by a moist winde," and neither were they and their adult the same species. This, she gathered, he'd taken from the ancients. But following his own claim never to have rendered anything he hadn't seen with his own eyes, she began to wonder whether spontaneous generation was nothing but a myth.

Marrel had left the family when Maria was twelve. Later she married his apprentice, Johann Andreas Graff, upon Graff's return from

---

* Merian referred to butterflies as "summer-birds."

an artistic journey to Venice and Rome. She was eighteen at the time, Graff twenty-nine, and soon enough she was pregnant. Shortly after her first daughter, Johanna, was born, they moved to his hometown, Nuremberg. The family home in the Milk Market neighborhood was once an inn, full of light and spacious. Down the road Johann Pachelbel would soon arrive to play the organ at the double-spired Lutheran church of St. Sebald's; up the hill was where Albrecht Dürer once lived, painting rhinoceroses and stag beetles. All day she would cook and clean and embroider and launder; she would feed and tutor and when she was spent, sleep. She painted, too, and whispered to herself, trembling, that her talent might be greater than her husband's. But neither the painter's guilds nor the art academies would admit women in those days.

Recently, a French judge had suggested that just as herbs and plants spawned worms and serpents, so might werewolves be transmuted forms of Satan. The Jesuit priest Athanasius Kircher claimed a dragon's

The Graff family home in Nuremberg, photo by the author, 2024.

head had been sent to him by a Romanian hunter, though it was too decomposed to display. All across Germany, people were burned at the stake as witches; there were public trials nearby in Würzburg, Bamberg, and Mainz. The fairer sex was time and again the one on whom eyes cast suspicion. Even the imperial mathematician, Johannes Kepler, was made long ago to defend his mother against charges of sorcery in Leonberg, accused as she was of appearing through closed doors and paralyzing a school master with a glass of wine.

Maria tried to align herself with powerful painters and engravers. One of them had been an apprentice to her father's father, and taught her half brother, Matthäus Merian the Younger. He included her in a list of German painters, remarking that she specialized in flower decoration, fruits and birds, alongside the excrement of worms, flies and spiders, and all such creatures, in all possible permutations. She began regularly teaching painting to a company of maidens in her home. Soon she published a book of flowers, *Blumenbuch*, after her grandfather's *Florilegium*.

But aside from the paintings, there were always the insects. The beetles and the flies and the ringlet summer-birds and gnats. As Kim Todd describes, she made room for them all in small jars and boxes, among her paintbrushes and etching knives, boiling kartoffeln, and hidden fantasies. There, in the many-windowed home, she performed small experiments, trying to see if heat could enliven a pupa, if she could pry open a caterpillar, discovering faint traces of wings. However crafty she was becoming with the paintbrush, she was not as dexterous with the surgeon's knife: When she cut a silkworm open, a colored, watery matter came out, oozing like a punctured boil with pus.

Ten years after Johanna, her first daughter, Maria was once again with child and wondered: If she pricked her belly, would it ooze? Or might it already be inside her, perfection in miniature, all assembled from the origin of the world . . . ? She did not know the answer to these questions. But she knew well, men didn't trust women, to say nothing of women who played with insects. And so she stayed at home, counting the days until Dorothea came, quietly feeding her caterpillars.

6.

EX OVO OMNIA?

Outside the Milk Market of Nuremberg, the ground beneath people's feet was beginning to shake. From London to Paris to Rome to Titiopolis, men observing nature began to wonder whether all the ancient wisdom they'd inherited was nothing but a ruse. Curiously, it was a seemingly innocuous question that was responsible: How to make a fly?

The Jesuit polymath Athanasius Kircher claimed he had the answer. "Collect a number of fly cadavers and crush them slightly," he wrote in his whale of a book from 1665 about everything from tidal waves to dragon's teeth. "Put them on a brass plate and sprinkle the macerate with honey-water. Then expose the plate, as chemists do, to the low heat of ashes or of sand over coals, or even of horse dung; and you will see, under the magnifying power of the microscope, otherwise invisible worms, which then become winged, perceptible little flies, and increase in size to full-fledged specimens."

Kircher was a flamboyant fellow. An adventurer and polymath, he was brought to Rome from Germany shortly after Galileo's trial so that the Jesuits could tout a heavyweight of their own. Accepting the testimony of travelers and peddlers at face value, Kircher became famous for his cabinet of curiosities, the *Kircherianum*. Guests could enjoy drinks from a mechanically spewing crustacean while admiring mermaid bones alongside a brick from the Tower of Babel. Named professor of mathematics and Oriental languages at the Collegio Romano, Kircher spent much of his time trying to decipher Egyptian hieroglyphs, which he hoped would uncover the language given to Adam, thereby prying open the secrets of God. He studied antidotes and electromagnetism, bioluminescence and demons, fossils and armadillos (the progeny of a turtle and a porcupine, he thought). He rejected the widely held belief in the existence of giants, based on the argument that had Noah and his family been giants, they would have sunk the Ark. Athanasius Kircher was somewhat short, and his eyes were suspicious. René Descartes said of him that he was "more charlatan than scholar." Gottfried Wilhelm Leibniz was even more dismissive: "He understands nothing."

But there was a problem. The man who'd been lowered into the mouth of Mount Vesuvius to investigate volcanoes, who defended the existence of both dragons and races of subterranean men; the man who'd invented the world's first pop-up book, the magic lantern, and allegedly a "cat organ" (just press the keys, hitting down on a row of cat's tails, making them meow in pain in different registers); he who would show up in the novels of Umberto Ecco, seeming more outlandish than the fictional characters around him—what that man said about how to make a fly actually worked.

Kircher's rival Francesco Redi would never believe it. Physician to the grand duke of Tuscany and keeper of the Royal Pharmacy, Redi scorned the notion that inanimate matter could be impregnated by spirit, that if the stars were aligned, one could create life by cooking alchemical elements in water. Never mind that Redi's patrons were the Medici, who had sold their own antidotes before they went into the more lucrative business of lending money. Two hundred years before the word was invented, he considered himself a true scientist, besides a courtier and a poet.

To counter Kircher and his magic, Redi devised a series of impressively controlled experiments. What he showed, having cut up a host of ox, deer, buffalo, rabbit, tuna, lion, swordfish, eel, lamb, and snake meat, was that when he placed the putrefied meat in carefully sealed flasks and waited, he failed to find flies and maggots. These appeared only when the flasks were left uncovered. Redi refused to break with Aristotle altogether. He wasn't prepared absolutely to deny, for example, that wasps and beetles and other insects might be produced by rotting wood, in the knobby growths known as tree galls. But while professing his love to the ancient "princes of philosophy," he clarified that he did not wish to bind himself "to swear that all they have said or written is true." Insects were certainly generated by sex—he'd seen it with his own eyes. The reason Kircher (and Aristotle, and all the rest) thought flies appeared spontaneously from putrid flesh was that they hadn't observed carefully enough. "No animal of any kind is ever bred in dead flesh unless there is a previous egg-deposit," Redi proclaimed, thinking that he'd put spontaneous generation to rest.

Thanks to an earnest Dane who'd arrived in Rome and become his friend, Redi had an impressive ally in his experimental method. Niels Stensen, "Steno," as he was known in Latin, was the son of a goldsmith. He'd traveled to Amsterdam and as a mere apprentice made a name for himself by discovering the duct that brings saliva to the mouth (Steno's duct). He'd moved from there to Leiden, home to one of Europe's most radical medical schools, where the dry texts of the ancients were being replaced by live dissections in an amphitheater—the theories of Hippocrates, Aristotle, and Galen challenged by a new mechanistic spirit taking hold. When Steno graduated, he traveled to Paris, where King Louis XIV had recently endowed a new Académie des Sciences. The man who had convinced the French monarch that a national home for science was necessary was Melchisédech Thévenot, inventor of the spirit level. Thévenot was also the author of the popular book *The Art of Swimming*, an orientalist, traveler, bibliophile, and ex-diplomat and spy. It was at his home—his personal *académie*—that Steno delivered a lecture on the anatomy of the brain. And it was at Thévenot's suggestion that Steno traveled to Rome with a message for the city's men of science: The problem to crack now was generation.

Times were changing. In England, a group of natural philosophers calling themselves the "Invisible College" had received a charter from King Charles II for a Royal Society with the motto *Nullius in verba* (Take nobody's word for it), expressing the fellows' determination to question even established authorities and to verify all statements by an appeal to facts—as determined by experiment. In Florence, students of Galileo founded the Accademia del Cimento with financial support from the two richest men in Italy, under Dante's banner *Provando e riprovando* (Test and test again). Science was becoming institutionalized and morphing into a new kind of pursuit. Little more than a century before, the influential Swiss physician and alchemist known as Paracelsus had provided instructions for creating a tiny human being: Allow human semen to putrefy for four weeks in a warm sealed container, feed this thing human blood, and after forty weeks you'd have a miniature infant. For the experimenters in the new academies of science, this was no longer good enough.

Not that the new academies rejected all written authorities. John Wilkins, one of the founding members of the Royal Society, aimed to reconstruct the universal language humans and animals had all understood in the Garden of Eden, before man (and especially woman) had lapsed. After the Fall, human beings had lost the knowledge of creation that allowed Adam, in his innocence and plenitude, to name all creatures "according to their natures"; recovering that lost language would help science properly see the world. Others suggested that even before language, signing with hands captured the essence of reality; a Londoner by the name of John Bulwer invented the word *chirologia* (from the Greek *chiro*—hand, and *logos*, language) to describe the natural language of the hand, becoming an early proponent of education for the deaf and dumb. Rather than being incapable of learning because they couldn't hear or speak, deaf and dumb people communicated in the most natural of languages of the animate body—the eyes. In fact, because the eyes were sensitive to the tiniest nuance, humans had developed all kinds of distortions of the body in order to inhibit others from reading their true body language. Bulwer himself wrote a book on the matter in 1650 called *Anthropometamorphosis*. It's subtitle was: *Man Transform'd, or the Artificial Changeling. Historically presented, in the mad and cruel Gallantry, foolish Bravery, ridiculous Beauty, filthy Fineness, and loathesome Loveliness of most Nations, fashioning & altering their Bodies from the Mould intended by Nature.*

After putting his new friend Redi onto the problem of generation, Steno took up residence in Tuscany under the wing of the Medici grand duke. It was the spring of 1666, and it seemed to him that the question of where we come from had never been so urgent. Steno had read the bestseller by William Harvey, the "humorous but extremely precise" insomniac Englishman (as he'd one day be referred to), famous for showing how the heart pumps blood through the body. Like Aristotle, Harvey believed that insects arise by chance, rather than by sex. After all, they had neither internal organs, he claimed, nor "true" parents. But Harvey was adamant about correcting the great teacher on one particular matter. The frontispiece of his 1651 book *Generation of Animals* showed Zeus delivering all kinds of creatures from a giant egg, and his epigraph made the point clear. Whether in animals that give birth to live young or in those who lay eggs, the same truth holds: "*Ex ovo omnia*," it said there—all things come from an egg.

Now, what Harvey meant by an egg was not the contribution of the female to reproduction, Aristotle's second "semen" alongside the sperm of the male. Harvey's egg was the actual early fetus, which to him looked like an egg without a shell. As Physician in Ordinary to Charles I before the English Civil War, Harvey accompanied the king on royal hunts and was allowed to dissect does that had recently mated. As he explained in his *Generation of Animals*, the act of copulation left no visible trace of semen, menstrual blood, or egg in the uterus. Even what was then called the female "testicle" (the ovaries) showed no change in size or shape. When the egg finally did show up in the uterus, it seemed to appear out of thin air. Harvey therefore had to go against two thousand years of received wisdom: Whatever the contributions of female and male to generation during sex, the egg was clearly not a direct result of two "semens" intermingling.

Perhaps the sperm brought an immaterial "spirit" with it, like an odor or a spark of lightning. Perhaps it mysteriously awakened an egg in the womb, the way a thought is fomented in the brain. The female "testicle" clearly had nothing to do with it, but one thing was for sure: However it got there, whether in mammals, birds, or insects, life always came from an egg.

Steno was intrigued. Already in his lecture at Thévenot's home he'd struck a blow against the great Descartes. The recently deceased French philosopher had distinguished man from other animals by pointing at a unique feature: a vibrating pineal gland in the middle of the brain. This, he claimed, was what gave humans their rational capacities. But dissecting the brain carefully, Steno showed that if humans had a unique soul, its origins had to come from somewhere else: The pineal gland could not possibly vibrate because it is as immobile as a rock. In a posthumous supplement to the French edition of his *De Homine*, Descartes tried and spectacularly failed to provide a mathematical description of how heat acted on "particles" to form a living body. (Following the ancients, Descartes believed such particles could come from either decaying matter, as in spontaneously generating insects, or from two semens—male and female—mixing together.) Everywhere, empiricism was winning out over rationalism, the new experimental anatomy over mechanical theories propped up by math. And the question of generation was on many minds.

Doing better wasn't going to be easy, though. Having trained in hard-headed Leiden, Steno saw that even the careful surgeon Harvey seemed to be groping for answers. There had to be a more convincing explanation for the origin of eggs, but what was it? A year before the appearance of Redi's own treatise debunking spontaneous generation, in March 1667, Steno published a short essay titled "Study of the dissection of a dogfish." In this nine-page report, he dropped a bomb that would change our view about where we, and all other life-forms, come from. Finally, the different strands of "generation" were beginning to come undone.

What Steno wrote was this: "The testicles of women are analogous to the ovary"—funny sounding words to modern ears, but a giant revelation. He didn't show this in a detailed drawing, nor did he perform any kind of experiment. But he declared that he'd seen it with his own eyes, and finally understood why it was important: Steno had dissected all kinds of animals—sheep and humans and lizards and birds and sharks. They were different creatures, but his claim now was that they were identical in the way that mattered most: Whether giving birth to live young, or laying eggs to hatch, all female creatures, including humans, had ovaries that carried eggs. And just like in birds, where those eggs come down through an oviduct, in viviparous creatures the eggs travel from the ovaries to the uterus. The meaning of this was nothing short of momentous: Women didn't contribute "semen." They contributed an egg. And all life on Earth came from it.

If Harvey had been puzzled about what an egg was, Steno suffered no confusion—although as a religious man he would have been appalled by the idea that God wasn't necessary to bring about life. Soon he would convert from Lutheranism to Catholicism and leave science behind. Having been appointed vicar apostolic for the Nordic Missions, he became a leading figure in the Counter-Reformation, and was consecrated with the titillating title of titular bishop of Titiopolis. Adopting a life of extreme piety, Steno died a poor man in Germany in 1686, suffering and emaciated. Pope John Paul II beatified him in 1988.

And Harvey? We know today that he simply chose the wrong animal: Since male deer go into rut a month before females ovulate, there was no way he could have found an egg in a dissected uterus, nor see any changes yet in her "testicles." It was therefore easy for him to fall in with the common Adamic wisdom, that woman had come from man, rendering her "testicles" vestigial and moot. For him, fertilization bordered on the mysterious: Semen acted without touch, eggs appeared like thoughts in the mind. How precisely such miracles happened he couldn't say.

And yet Harvey also asked the question that had been on everyone's mind since Aristotle: Whatever females and males bring with them, how does a life assume its particular form? Aristotle argued that living matter had a potential that was awakened by the male semen, providing an external cue that was automatic and as determined as the moving hands of a clock. That's why when he cut open chick embryos at different stages, he always glimpsed the heart appearing on the third day, detected all the body parts after exactly ten days, and could see a coat of feathers by the twentieth. Why things developed in that order he called a chicken's "final cause," a term that simply meant that the goal of the chick embryo was to become a chick, and that it proceeded accordingly. But Harvey wanted to rid natural philosophy of this kind of argument—an explanation we now call teleological (from *telos*, or "purpose")—which on closer inspection, since devoid of any mechanism, seems to explain nothing at all. To describe what happened after fertilization, he coined a new word—*epigenesis*—defining it as the addition of parts budding out of each other as the egg grows. A hardheaded thinker who had managed to acquit four women accused of witchcraft, Harvey knew that the heart and tissues and organs that mysteriously appeared had to come from somewhere, no magic involved. And so Harvey said the egg contained within it, already fully formed, a representation of the creature it would become. Thinking of genes, the historian Matthew Cobb writes that Harvey's explanation was "both tantalizingly modern and infuriatingly vague." Because how this happened, or how it could be observed and studied, Harvey had no clue at all.

Still, Harvey had great respect for Aristotle, adopting his idea of "imperfect" and "perfect" animals. The larvae and the adults of bees and beetles and butterflies and wasps weren't two distinct stages of the same life-form, he wrote, but two distinct creatures that had both arisen spontaneously from decay. No epigenesis for them, no internal representation and gradual self-assembly. No, such creatures developed in a different way altogether: all at once, like a seal being stamped on wax. With his own eyes he'd seen wings beneath the pupal case just moments after it was created. To give this a name, Harvey took an old word, and contrasted it with his new word, *epigenesis*. Imperfect creatures aren't made like the rest of us, he counseled.

Instead they undergo *metamorphosis*.

7.

AN INGENIOUS WOMAN

Back in the Milk Market of Nuremberg, word got out. Aware of her passion, friends sent Maria Sibylla Merian caterpillars as gifts, and the curious came by asking to see her insects. Meanwhile, she kept on drawing flowers. On occasion she joined the meetings of The Order of the Flowers on the Pegnitz, a group of men and a few women who loved nature, poetry, and the German language in equal parts. They took code names—Hedge Rose, Meadow Clover, Mayflower—went on walks, and read each other's rhymes. It was pleasant. But Maria knew about the dangers of beauty. How men and women had ransomed their lives in Holland to put their hands on a tulip. And she began to notice, too, that whereas they had once been mere decorations to the flowers in her *Blumenbuch* book, insects were now commanding her full attention.

A member of the Order, a professor of languages and poetry, had traveled far and wide. From him Maria heard of a man named Robert Hooke, in London, placing a piece of cork under a microscope and calling the catacombs he saw "cells." And of the doctor from Bologna, Marcello Malpighi, whose careful studies of the date pits* of silk-worms found faint traces of antennae, wings, and legs. The shapes dissolved into liquid when he pierced them, but she was listening, intently. To tales, too, of the master of miniature of Amsterdam, Jan Swammerdam, dissecting insects with great dexterity. It was Swammerdam's *Natural History of Insects* from 1669 that affirmed her observation that all insects come from an egg laid by an insect of the same kind.

Still, the finest minds of the day could not explain how worms appeared miraculously in tree galls, or in the guts of dogs and humans. Nor did they know why cocoons often hatched flies from their pupae rather than moths. Despite that man from Rome, Redi, spontaneous generation was alive, as in ancient times. And even Maria wasn't certain that it could be safely put to rest.

The natural history books she knew were either about plants or about insects, never both. And the ones about insects never showed the different life stages on the same page, clearly not fathoming metamorphosis. That's why she loved Johannes Goedaert, the man from Middelburg. He was a still-life painter entranced by insects, like her, and sought to bring them into the light for the glory of God. Acting in this way, he put the whole life cycle of over one hundred species together—from caterpillar to date pit to imago—for the very first time. She owed him a great debt. And yet Goedaert's paintings were colorful but dead, since he had made an even greater omission. Far from the creeks and gardens, the man from Middelburg neglected on most occasions to include with the insect their eggs, and never the plant on which it lived and fed.

Maria wanted to capture Nature as she really was, to study not just shape, but movement and habits. Hadn't Pliny said that Nature can be seen nowhere more than in her smallest creations? So she went to the fields to learn from the caterpillars. There she found them, eating a certain plant and not another, preyed upon by this and not

---

* A term Merian often used for pupa.

that predator, flying, crawling, rolling leaves as shelter, attaching silk threads, interacting, living in a web. She took pains to write down all such details, spared nothing of her talents. One day we would give this science a name—*ecology*. But now Goedaert wrote that a moth drawn to fire was a warning to those seeking divine secrets who would likewise be swallowed by the radiance of God. Nature was God's handmaiden, to be sure, but Maria had other concerns. At night she dreamed about painting with color the theater of such miniature dramas.

Her second daughter was born on February 3, 1678. She named her Dorothea after her pupil Dorothea Maria Auer, a member of her "company of maidens"* and now a painter. Dorothea's brother-in-law owned a celebrated garden just outside the city walls above Maria's house, and she arranged that Maria be allowed to use it. The great uncle of Maria's other beloved pupil, Clara Regina Imhoff, was a principle guardian of the Imperial Castle. He gave her access to a spacious garden just two streets from her home, as well.

She wrote her flowers book in German for an audience of young women who knew no Latin. It sold handsomely, in three installments of twelve etched flowers each. Encouraged by their success, she and Graff reissued the earlier plates to put together all thirty-six. She wrote these words there, in the preface to *Neues Blumenbuch*:

> *Thus must Art and Nature meet with fond embrace*
> *And each one to the other extend her hand in peace:*
> *Who battles thus does well! For after such a duel,*
> *When all is finally done, both are satisfied.*

Maria could feel it in her gut. More and more the insects were calling her away from the flowers. And more and more she retreated from the company of people to consort with them, as they were closer to her heart. She roamed as far as Altdorf one day, twenty-five kilometers from Nuremberg, where she stumbled upon hundreds of

---

* *jungfern Combanny* in German.

caterpillars in the grass of an old moat. She spared no effort to find the caterpillars of the large ogled-eyed emperor moth, and the small and purple-hued pease blossom moth, recording the damage they made to leaves and buds, surely of interest to gardeners. She would search in the gardens and pasturelands and meadows, under bridges and beside small brooks. She noticed things. How satin moth caterpillars pined for the sun while others sought shade. Or the looping movement of the heads of peppered moth caterpillars, which gave her much delight. When she came out after dark, wrapped in a coat, she observed cutworm caterpillars emerging to feed from the soil where they had burrowed. She saw more than once the red beetles on the red willows that grow along the river, their eggs piled up together as if for a game of bowling. And this too brought much happiness to her heart.

Back at home in the attic, she put her treasures in wooden boxes with holes. It was hard work husbanding them until they pupated and hatched, from April through September. Some took a fortnight, others thirty days, the gypsy moth three full moons of patience. Some would eat, each day, three times their body weight, and Maria had to return to the fields to fetch foods for their finicky guts. If she failed to bring them nourishment in time, some would eat one another, others shrivel up and perish. No less careful was she cleaning their boxes, from dried-out plants, insects, and frass.* The frass was soil for fungi that would kill them, if the parasitoids didn't get there first. In May she would have at least fifteen boxes to tend to. But she did it always with love.

Many died, more than survived. Even in our day, in fancy laboratories, scientists can only bring to full cycle one of three of her small tortoiseshell and one of nine of her peacock butterflies. When she herself failed, she grimaced, resigning to wait for next season. It took her four years to complete the life cycle of the cabbage and dot moths, three years for the emperor. But when metamorphosis did stir in her attic, finally, she dropped all else. Then she would rush for her pigments, her brushes and vellum. Sometimes in the dead of night.

---

* Insect feces.

She gazed warily. It took the peacock caterpillar four to five hours to build its fig pit, and when it hatched, the imago's wings needed thirty minutes to unfurl. Once the adult was born, she could take her time to paint it. When she was ready, a hot needle would do the Lord's work quickly. Then she could pin it down and spread its wings.

The sketches in the notebook that would end up in St. Petersburg* were the foundation: At last Maria could consult her cobbled notes to fashion the final rendition. She'd begin with a clean copperplate covered with wax mixed with resin. Then, using the notebook as guide, draw the design life-size into the ground with an etching needle. Next she'd apply vinegar, to bite into the metal wherever the ground was removed by the needle. Sometimes, when the lines were blurred, she'd use an engraver's burin to sharpen them. Finally, she'd ink the copperplate and place it in the roller press, operating it by hand to produce the prints on paper. And while the ink was still fresh, she would place a new sheet of vellum upon the first print and press it again. Mirror-image counterproofs doubled her earnings.

Colored and uncolored versions of her book abound, and on the market the colored were dearer. Having painted, engraved, and etched everything, she had no need for the services of others besides a text printer. It was a good system, reducing costs, but would injure her posterity. Most bought uncolored books, and those who painted within didn't always heed her instructions. Tricked to believe that she was responsible, naturalists would sully her name for having rendered the wrong shade of yellow, cobalt, or red.

Still, Maria's heart fluttered. At last, the year following her second daughter's arrival, her *Raupen*** book signaled a further birth in the family. Her second daughter she named Dorothea, but this third book got a longer title: *The Caterpillar's Wondrous Metamorphosis*

---

* Scholars now believe the *studienbuch* that resides in the Library of the Academy of Sciences in St. Petersburg was recopied and rearranged by hand by Merian herself between 1686–1691. It contains 133 pages of paper in her handwriting and 338 pieces of vellum with her painted studies.

** German for "caterpillar."

and Particular Nourishment from Flowers in which for the benefit of explorers of nature, art painters and lovers of gardens through a completely new invention the origin, food, and development of caterpillars, worms, summer-birds, moths, flies and other such creatures, including their times and characteristics are diligently studied, briefly described from nature, painted, engraved in copper and published by Maria Sibylla Graff herself, daughter of Matthäus Merian the Elder.

There they were, laid bare, the life cycles of creatures, with the plants they ate and sometimes what ate them, all on one page. Look and see: Maria showed caterpillars devouring a plant, moths laying eggs, a summer-bird landing on a blossom. Nature as she breathed, just as God intended. Her learned friend from the Order, Cristoph Arnold, included a poem in the introduction that made her cheeks redden: "What Swammerdam promises / what Harvey once lost / now comes to the knowledge of all / that an ingenious woman has achieved all this herself / to occupy her time."

Plate 5 from Merian's 1679 *Raupen* book depicts a tiger moth and parasitic wasps around a hyacinth.

She wasn't sure yet that flies were not sometimes born spontaneously. She did not know why cocoons sometimes produced them instead of moths. Most of all, how a flying adult emerged from a fig pit spawned by a caterpillar remained a mystery. Lacking more hours, she wrote in her book that she leaves such conundrums to "gentlemen scholars," and went about her housework.

But in Maria's heart there was never more passion to know the truth.

8.

GENTLEMEN

Jan Swammerdam was hooked. His father wanted him to become a clergyman, but Jan was drawn less to the wonders of heaven than to those on Earth, like the specimens displayed in his father's own cabinet of curiosities. When the time came, he decided to go to medical school in Leiden, where experimental science was flourishing. Famous professors there had shown how blood circulated, how food and air went in and out, sustaining the machine that was the body. It was all so very exciting. Soon he too picked up the drafting pencil and dissecting knife. One day he'd be called the "master of miniature," and one of the great biologists of his age.

Reinier de Graaf was a little younger, and Catholic. He'd come to Leiden from nearby Schoonhoven with a brain matched only by his steady hand and ambition. Attending anatomy demonstrations in Leiden's new, purpose-built amphitheater, de Graaf, Swammerdam, and Steno became friends and peers. All three showed brilliance wielding the knife and using a single-lens microscope. Like Steno, de Graaf and Swammerdam each traveled to Paris to meet Melchisédech Thévenot. At the feet of the ex-spy and inventor of the spirit level, all three were persuaded that the greatest mystery to solve was generation. And so, while Steno was in the court of the Medicis writing his historic "Study of the dissection of a dogfish" about ovaries and eggs, Swammerdam returned to Leiden, where he finished a dissertation on respiration and turned to the question of how insects were born.* As a Catholic, de Graaf was ineligible to teach at Leiden; he settled in Delft, established a medical practice, and, about the time his neighbor Vermeer finished *Girl with the Pearl Earring*, began to study male genitalia.

It was about then, at the start of 1667, that Swammerdam received a message from his Leiden anatomy teacher, a man by the name of Johannes van Horne. Van Horne had a female cadaver in his possession and needed help: He'd been studying the connection between the ovaries and the uterus on the assumption that the female "testicle" produced an egg. Since books circulated slowly then, neither man knew that Steno had already made that very same suggestion. But van Horne did know that in frogs and birds the egg traveled through the oviduct, and his hunch was that human eggs went through the fallopian tubes. No one was a better draftsman than Swammerdam, and no one more adept at making dissections visible by a technique that involved injecting hot colored wax into blood vessels. Desperate for his student's expertise, van Horne invited him to assist.

A year went by, and the Medici prince Cosimo arrived in Leiden. Proud and composed, he handed van Horne the book by Steno in which appeared his "Study of the dissection of a dogfish." Because van Horne had failed to publish his own treatise, with Swammerdam's drawings, they'd been scooped. Privately, Swammerdam scolded his teacher. He also wrote to Steno from Amsterdam to suggest that while their teacher had tarried, he'd probably had the idea first. The newly converted Steno replied piously, with modesty and grace.

But de Graaf was another story. Having heard through the grapevine that his former teacher van Horne was working on the female

* Swammerdam had discovered that snails are hermaphrodites. The frontispiece of his classic dissertation on respiration sported two snails copulating, showing that his interests already lay in generation.

"testicle," he penned a quick abstract, or what was called a *prodromus*, of his own future book on the subject, in the form of a letter to another Leiden teacher named Franciscus Sylvius. De Graaf was hoping to establish his priority, but when van Horne got hold of the *prodromus* he was apoplectic. In a short public letter of his own, he claimed to have carried out his studies before de Graaf, and without knowledge of his work. De Graaf, he hinted, had stolen his discovery. As the argument devolved into an ugly spat, van Horne got sick. He died, aged forty-eight, without publishing his great finding of the female egg.

Freed of his nemesis, de Graaf went ahead and published his "Treatise concerning the generative organs of men; on enemas and on the use of syringes in anatomy" in the spring of 1668. There, with commanding skill, he exposed the shortcomings of the great Aristotle: No, testicles weren't a solid gland, but rather made up of densely packed tubules. They weren't weights that merely helped the semen along. They actually produced it—a fact somehow lost on Aristotle even though castrates were known to be sterile. Aristotle had been remarkably unfair to women. Without women there would be no next generation, and it was an insult, de Graaf wrote, to suggest that a woman was merely "an incomplete male." Unfortunately, when it came to the actual workings of procreation, de Graaf adopted a distinctly Aristotelian worldview: Male semen was produced by blood due to the action of "animal spirits." And it was that semen, and that semen alone, that was capable of generating a soul. When it reached its destination, it "fermented" the menstrual blood. And generating a fetus, it would slowly begin to grow.

As European men competed to crack the mystery of where humans came from, they did not neglect the generation of the many-legged, the hairy, and the small. Three years before de Graaf published his treatise, Swammerdam put on a remarkable demonstration for Thévenot and the other learned members of the Paris *académie*: Opening a small box, he took out a smooth, creamy silkworm caterpillar that was about to pupate. Pinning it onto a prepared pad beneath a microscope, Swammerdam then made a cut from just below the head down the length of its body. And to everyone's amazement, there appeared beneath the skin, as if in a magic trick, the outlines of its future head and wings. When the "master of miniature" finally pierced the watery structures in front of his mesmerized audiences, they dissolved, but the message was clear: Contrary to the teachings of everyone from Aristotle to Johannes Goedaert, butterflies and caterpillars were, in fact, the same creature. As Swammerdam put it, the pupating caterpillar didn't just contain "all the parts of the future animal, but is indeed that animal itself."

It may not sound like much to modern ears, but the implications were momentous. First, Harvey had been wrong when he repurposed the word *metamorphosis* from its ancient origins: Since the rudiments of the adult form were in place in the juvenile, it was simply untrue that insects developed from decay all at once, as if stamped in shapeless wax. But the second implication was even more wild. When Thévenot sent Swammerdam a book called *The Silkworm* in which the Italian physician Marcello Malpighi carefully illustrated and explained what had become Swammerdam's own party trick, he knew now for certain that he had to be right. There were no such things as "perfect" and "imperfect" creatures. There was nothing unusual, or mysterious, about insects at all! As Swammerdam went on to show, they were of a piece with tadpoles and bishops. In all life-forms, including humans, like breeds like. It happens gradually, by epigenesis, and always from a female's egg.

But how? In his book *Natural History of Insects*, which made an impression on Maria Sibylla Merian, Swammerdam showed convincingly that the age-old conundrums of galls and parasitoids were nothing but fluff. Had all the great naturalists who'd come before him, including Redi, observed more keenly, they would have seen that wherever a creature came from, another female of its kind had already, invariably, been there with an egg. But when it came to how epigenesis actually worked—how the slow "swelling, sprouting, protruding, and the appearance of new limbs" came about—Swammerdam could only guess. Harvey's ideas on the matter were no help, he wrote, containing "almost as many errors as words."

Swammerdam was a god-fearing man. The idea that creatures might come into the world spontaneously smacked to him of chaos, disorderliness, the precise opposite of divine intent. By pointing to the regularity of things, he hoped to make manifest "the stupendous works of the most adored and all-wise creator." In point of fact, Swammerdam didn't believe in generation, as such. Instead he believed in propagation, the chance-free process of growth of parts. For him, all future men and women were already present in Adam and Eve, complete with their original sin. What it meant was that, without exception, God had created all of us. And that humans would cease to exist when the last egg was spent.

Soon Swammerdam and his friends would find themselves involved in another feud. A younger graduate of Leiden named Theodor Kerckring published a pamphlet in which he claimed that eggs in the female "testicle," or ovary, could be found in married women as well as in virgins. Hearing this news, Swammerdam immediately wrote to his old friend de Graaf with a warning; he knew de Graaf was working on a new book about the genital organs of women, and exhorted him to publish before he, too, was scooped. At first de Graaf ignored him. Only when he caught wind (from Swammerdam) that Swammerdam had shared his drawings of the uterus and ovaries with yet a fourth researcher,* was de Graaf moved to act. Once more putting out a *prodromus* to stake his claim, de Graaf showed for the first time the fine structure of the clitoris, and argued that the eggs in the follicles traveled down to the uterus through the fallopian tubes. And, no less importantly, he expressed his worry that those to whom Swammerdam had shown his drawings might "gain credit for my work." When he finally published what became his masterpiece, called *New Treatise Concerning the Generative Organs of Women* in 1672, he was able to show with the most realistic drawings to date that the clitoris was the source of female pleasure. Clearly, pleasure had been designed as an aid to procreation, both in women and men. More than that, by injecting water into a uterus, de Graaf showed that

* The Dutch physician and anatomist Gerard Blaes (1627–1682).

the only route in and out of its top was through the fallopian tubes. By dissecting rabbits, rather than deer, as Harvey did, he argued as forcefully as he could without actually observing it, that eggs travel from the ovarian follicles through the fallopian tubes. Unfortunately, since mating induces ovulation in rabbits, who are not on a monthly cycle, de Graaf never properly understood human menstruation. Still, he was the first to prick a human follicle, and see the thin liquid oozing out. It looked like the white albumin of a chicken, and when boiled it had the same consistency and taste. De Graaf now changed his mind: The male semen ferments the egg rather than the menstrual blood. He was happy to acknowledge the dead (and therefore unthreatening) van Horne, but snubbed Swammerdam. Deeply offended that his former friend might have tried to upstage him, he didn't even mention his name.

Judged by today's standards of scholarship, it's fair to say that it was van Horne who first suggested that females have eggs in their ovaries, and that these travel through the fallopian tubes to the uterus; that Steno first stated this in print, generalizing it to all creatures that give birth to live young; that Swammerdam, called upon by his teacher, first clearly illustrated the ovaries and follicles; and that de Graaf was first to prove, based on experimental evidence, that eggs are released from the ovary. At the time, however, the Royal Society determined that the winner in this feud was none other than Steno. It was his "Study of the dissection of a dogfish" from 1667, they claimed, that first unveiled the existence of a new vision of generations in all animals. De Graaf never lived to see the officials call the race: He died a week before the appearance of the Royal Society's decision at just thirty-two years of age. After falling briefly under the spell of the itinerant Flemish preacher Antoinette Bourignon, Swammerdam devoted his final years to an epic and very devout treatise on insects. Using a microscope and a dissection knife, he described countless tiny creatures, from mayflies to stag beetles to cheese mites. It was his description of the eponymous lowest of the rank that went on to be a classic: "Herewith I offer you," he wrote in a book published posthumously called *The Bible of Nature*, "the Omnipotent Finger of God in the anatomy of a louse." He died in 1680, age forty-three, having never married, or, as far as we know, gotten anyone pregnant. Since there is no reason to think he ever encountered the work or name of Maria Sibylla Merian, he could not have known before he passed that she was about to change her life dramatically. Just a day's sail down the Rhine, Maria was preparing.

9.

WHITE WITCH

When her stepfather Marrel died in 1661, Maria and her husband and children returned to Frankfurt to care for her mother. There, to his credit, her husband helped her publish her second *Raupen* book. But she and the world around her were in a state of agitation. Despite the peace that followed the Thirty Years War, the distance between the rich and poor had never been so great. Fashions from Paris walked the streets alongside the sick and maimed. Those upon whom fortune had smiled thought themselves deservedly higher, above the wretched. Some, disgusted by these displays of pride, retreated to the countryside. Others, like the Quakers, abandoned Europe altogether. When the pietists began preaching, many flocked to them. One need not be a priest, they taught, to speak directly to God.

Maria's own half brother, Caspar Merian, had joined the Labadists in Holland. From him she heard the tale of "the Star of Utrecht," Anna Maria van Schurman, who was supposed to have written an Ethiopic dictionary, who played several musical instruments, wrote poetry and essays, and engraved in glass and painted. In one of her tracts, *The Learned Maid*, she called for women to study history and mathematics. Yet now she had given up her studies and joined the sect. She was not alone. Many learned women were drawn irresistibly to a creed that would allow them to speak with God directly, without the mediation of men.

For her part, Maria plunged ever deeper into the mysteries of insects. Though their transformations were divine miracles, she continued to ponder those that seemed unnatural, and morally suspect: flies that injected their eggs into the larvae of moths, for example, using husks of the host's head for protection during their own metamorphosis. Even worse were those who invaded other species' pupae, gorging on them from the inside. One day we would call such perversions "parasitoids," but for her they were provocations: How could a beneficent God allow such evil in His world?

At thirty-nine she finally left the home of her husband. We have no hint of the cause. What was she to do? Where was she to go? She visited her father's grave, but instead of solace she records finding a pale caterpillar and tree gall.

For reasons we can only guess at, she took shelter with the Labadists. And in the small Friesland town of Wiuwert, together with her aging mother and her two daughters, Johanna and Dorothea, seventeen and seven, she donned the rough wool habits behind castle walls. She removed her jewelry, slept on a narrow bed, kneeled on cold stone floors for dawn prayers. Driven away by the Catholics and revered as the "second Calvin," the founder of the community, Jean de Labadie, had died years earlier in Altona, but his teachings lived on: All Christians are elect; men and women are equal; the Church is not of this world; know God through prayer and mystical devotion; Augustinian predestination, self-denial, and fasting are good for the soul. Turning inward, slowly, Maria grew accustomed to the rhythms of her new world.

Anna Maria van Schurman had died before she arrived, but there was a printing press selling her and Labadie's inspirational teachings. There was a doctor named Hendrik van Deventer, too, concocting pills for the ill, and "Labadist wool," which enjoyed a local reputation. As a probationer, the equivalent of an apprentice, she was assigned chores cooking in the kitchen and cleaning clothes in the washbasin. And yet,

Kim Todd tells us, without touching them for two years, Maria kept hidden her books and notes, her paints and brushes and vellum, the *carta non nata*—skin from unborn lambs—which held color better than all the rest. Pietism had its virtues, but she found reasons for the indiscretion. As gentlemen before her had taught, pace Anna Maria van Schurman, God could be discerned in the structure of a louse.

Graff arrived unannounced one day, demanding from behind the walls that she return to him. It was wrong for a woman to leave her husband, he shouted, but she rejected him nonetheless. The Labadists did not let him in, and his daughters refused to see him, even when he got sick and grew weak outside the castle gate. There, sitting on a wooden stool, he sketched the compound and the two lines of poplars that led to it. Maria took pity on him, and visited him, but their differences were too great. One morning he gave up and left, and she never laid eyes on him again.

Instead, she feasted her eyes on the natural wonders that made their way to desolate Friesland: giant moths and colored lizards, sparkling beetles and exotic fruit. She was even shown a twenty-three-foot anaconda, a strange-looking lanternfly, and a pineapple. Waltha Castle, after all, was owned by Cornelis van Sommelsdijck, the first governor of Surinam, across the oceans. Men said he'd imported horses there from New England, dug a canal between the Surinam and Saramacca Rivers, imported peach trees and ginger, built a town hall. His twenty-two-year-old sister Lucia had married Labadie when he was in his sixties, and lived in Wiuwert long after he died. When Sommelsdijck left for Surinam in 1883, he took her and other Labadists with him. Later, another sister joined, helping to build a colony upriver, in a place called La Providence. The poor governor, they were told, was murdered by his own soldiers, and, one after the other, the Labadists and their children in the jungle got sick. Those who returned alive whispered secret curses against their leaders. But they also brought with them the jewels of the New World.

Inspired, Maria returned to her caterpillars. Watching moth after moth emerge from its fig pit, she dreamed of a world beyond the walls of Wiuwert. It was well-known that Sophia of Hanover, the mother of King George I of Great Britain, had visited the colony. That William Penn the Quaker had been there, judging of the pious community that "certainly, the Lord hath been among them." Even the great philosopher John Locke bowed his head, passing through the Waltha gate. Still, her eldest Johanna had turned twenty-two and was single. And Amsterdam was less than a day's ride away on horse, teeming with markets of flowers and spices and men. All rivers and roads and canals lead to *Le grand monde*, people said. They brought paper from the Orient and narwhal horns from the north; coffee and sugar and cocoa from the south, manatees and rattlesnakes and dodos. God's workings were being unmasked, from the greatest to the smallest. Just upriver, in Delft, Antonie von Leeuwenhoek used a microscope to chart the life cycle of a flea. And across the channel in Cambridge, a magus named Newton had penned the laws of the universe.

Maria's mother died in 1690, and that summer, disease crept into the castle, killing one after the next. Gradually, the sect disbanded, like a caterpillar in a cocoon. Doctor Deventer cried that the "community of goods" was nothing but an injustice: It was he who concocted the pills, should it not be he who gained the wealth? The sect's disgruntled translator, she heard, was preparing a book titled *Rise and Fall of the Labadists.* If God existed, many felt he had abandoned gloomy Wiuwert. Electrified but altogether numb, Maria shed the coarse wool habits in haste. Packing her effects, she grabbed her daughters, and left.

Some perhaps judged her flippant, but in the end *Le grand monde* proved a grand distraction. Amsterdam pulled her in, but after eight years she was desperate to get out. At least she had reclaimed her independence, and her name, Maria Sibylla Merian. She had socialized, sold paintings, dined gracefully, and read many books. She had met the high and mighty: fellows of the Royal Society, the oil-painter daughter, Rachel Ruysch—an equal to Rembrandt some said—and her cabinet-of-curiosities-making father. But all the while, her investigations of metamorphosis stalled, and the comfort of things dried her throat. She was fifty-two years old, and may not have further occasion. At the home of the burgomaster of the city, she'd seen wondrous insects from the West Indies. And so she placed an advertisement in the newspaper one day: all 255 of her paintings for sale. When a buyer was secured, she wrote a will and prepared herself. It was June 1699. The next day, Dorothea and Maria boarded ship to Surinam.

With the Lord's providence, through fierce gales and cold, she avoided pirates, and, as we've already learned, settled into a three-story whitewood house the Dutch had built in the capital Paramaribo. The sun was merciless and the streets made of crushed seashells. She took some slaves, and an Indian servant. Not that she was blind to excesses: men and women dragged in shackles, limbless youth cruelly punished for minute infractions, others caught trying to escape, hung outside the fort on metal hooks. But Maria showed charity, and kept her mind on insects. On other creatures, too, like the scarlet ibis, called the "Surinamese flamingo," and on pineapples, which she rendered in watercolors, growing around her house like weeds.

In the spring of 1700, Dorothea and Maria traveled upriver to La Providence. Only a few of the Wiuwert sect remained, and they were bitter and confused. They had come searching for God, but now the Devil's blood coursed in them. Some nights she sat with them in silence. Had she ever been one of them? Had she ever truly been with God? We don't know what she felt. But we might imagine her tossing and turning in her cot at night, listening to the song of cicadas.

Outside, the forest was alive: coatis, frogs, howler monkeys, aqua cotingas, and untold caterpillars, moths, and summer birds. Each morning Maria's slaves would help her hack through the thicket to capture little jewels, sparkling with *glanz*.* When she found eggs, she placed them in a leaf-padded box and carefully carried them back to the barracks. One ghost moth emerged one day, a "white witch" tinged with violet with soft beige and brown markings. Its wingspan was the largest of all New World moths—almost a foot long—and at that moment she judged herself the luckiest woman on earth.

---

* A Dutch word Merian used repeatedly to describe the sheen or gloss of insects she found so hard to paint.

Her luck ran out, though, thanks to a fever that turned her skin yellow and her eyes red. The fever affected her brain as well, clouding her keen intellect. She had planned to stay three years more, but it was time to leave now. With her on the ship *Peace*, besides Dorothea and an Amerindian servant, were countless boxes filled with geckos and beetles and palm weevils and sphinx moths. Most valuable were the silk cocoons, small turtles, and iguanas. A shame the pearly lizard eggs hatched on route: With no mother present, all the tiny dragons perished.

Back in *Le grand monde*, Maria went about toiling on the book, the grandest she would author. With the help of the apothecary and butterfly collector James Petiver, she published it, finally, in 1705. Exotic species expert Caspar Commelin showed her grace, translating her Dutch to Latin. Readers marveled at the color plates showing lanternflies and cicadas, pineapples and cassavas, skippers and grasshoppers, morphos and swallowtails, and tumble bugs and caligos—sixty folio-sized in number. There, she showed the reproduction and transformation of animals, how one emerged from the other, and the properties of their foods. Her talents, she might have admitted, were far outstripped by her ambitions. But she was satisfied to read the reviewers praising the "great passion for investigation and tireless diligence of this woman."

Maria had heard by then of the battles between the long-deceased Jan Swammerdam and von Leeuwenhoek, still gazing down the microscope in Delft. How the first believed it was the female who contributed the germs of potential, whereas the second swore it was the male. Had Adam, or Eve, contained within them all future humans? And was it the same for butterflies? She could not say, but one thing she was sure of: When wasps emerge from caterpillar fig pits, it is not by spontaneous generation but because their mothers had injected their eggs there, and tree galls are the same. There was good and evil in nature, she told herself.

Both were part and parcel of God's world.

10.

NABOKOV

Maria Sibylla Merian died in 1717. She had lived a life of form and content, hardly "imperfect" in any sense. There had been molts and transformations aplenty: a daughter of a master, an artisan wife, a mother, a pietist, a metropolitan artist, collector, and businesswoman, fearless explorer in the tropics. In the engraving made of her by her son-in-law in the year of her death, her eyes show a weary intelligence as her mouth hints at a smile. Life hadn't been easy for her. But what a life.

In the future, great minds would remember her: The English naturalist Erasmus Darwin, grandfather to Charles, cited her in his books; the German polymath Goethe marveled at her blending of art and science; Linnaeus consulted her studies in the naming of 136 species of plants and butterflies. After the study of "biology" (a term coined in 1802 by Jean-Baptiste Lamarck) became professionalized, her reputation suffered for a time, in part due to forgers who'd colored in her drawings incorrectly and tried to pass them as originals. It was feminists of the early twentieth century who first retraced her influence. Eventually, Germany would put her face on its currency; the United States adapted two of her Surinam plates into postage stamps. Even

Engraving after a portrait of Maria Sibylla Merian by her son-in-law, George Gsell, 1717.

the literary luminary Vladimir Nabokov was profoundly influenced by Merian's life.

As a boy, Nabokov discovered her Surinam book in the attic. The world of metamorphosing creatures was utterly mystifying. For seven years he kept a hawkmoth cocoon dormant in a box, until it hatched one day in the heated compartment of a train he was riding. Nabokov became an entomologist, specifically a lepidopterist, an expert on butterflies. He later considered his work in the 1940s on the genitalia of American blue butterflies "the most delightful and thrilling in all my adult life," his "immersion in the wondrous crystalline world of the microscope, where silence reigns . . . so enticing that I cannot describe it." When his eyesight began to fail, Nabokov turned to his pen and away from the ocular. But he remained a careful observer. His literary critics would accuse him of obsessive attention to detail, "more interested in the subspecies and subgenus," Nabokov put it jokingly, "than in the genus and the family."

Thanks to Redi and Malpighi and Swammerdam, insects were no longer considered natural symbols of Christ, nor thought to sprout miraculously from mud or fire. Maria had been born into a world from which, to this degree, the supernatural had been expunged. Yet she and they were still God-fearing people, who believed in the orderliness of His creation and who thought this order would be laid bare step by step, one discovery at a time. They had no idea how the paths of scientific knowledge could twist and turn.

During her own lifetime, a stark shift in perspective occurred when Antonie van Leeuwenhoek, a well-to-do draper and autodidact, made love to his wife one day in the autumn of 1677 in Delft. Less than "six beats of a pulse" after ejaculating, he scooped up his semen, sucked it into a capillary tube, put the tube under the microscope, and pressed his eye to the lens. Leeuwenhoek's lenses could magnify objects up to five hundred times larger to view objects as small as one thousandth of a millimeter—much greater resolution than anyone before him had ever achieved. And on that autumn day,

Vladimir Nabokov collecting butterflies toward the end of his life, in Switzerland (1970), photograph by Horst Tappel.

as he described it, there suddenly appeared "vast numbers of living animalcules" under the microscope, that "moved forward with a snakelike motion of the tail, as eels do when swimming in water." Eager to share his discovery with the world, but also keen to make the whole affair seem less obscene, he had his crude Dutch translated into high Latin before sending his letter to the Royal Society, begging that "it be regarded as private." Having found such swimmers in codfish, cats, frogs, and even fleas—a million, he calculated, in a volume equivalent to a coarse grain of sand—Leeuwenhoek was now certain: Harvey and Steno and de Graaf and Swammerdam had all been mistaken. It wasn't the female egg that gave life, but male semen instead. In humans and other mammals, they swim up the fallopian tubes, attach to veins and feed on them, shed their tails, and merge into a ball. That was the supposed "egg" that everyone claimed to have found—not any part of a female, but a little animal from the male with a living soul.*

So began a war between "spermists" and "ovists" that lasted nearly two centuries. From our modern perspective it seems strange that female and male wouldn't have been recognized as two complementary components. But when it came to the question of where life comes from, one side held firmly that sperm just awakens the egg, whereas the other claimed that eggs are mere nourishment for sperm. Perhaps, as the historian Matthew Cobb has suggested, in an age when mechanical clocks were used as analogies for living creatures, it was hard to see how two clocks could be dismantled to make a third. Today we think in terms of genes, and of traits inherited from both parents, but the problem then was that no one yet understood heredity, and many remained unconvinced that there was anything of the

* No one actually saw an egg until the nineteenth century. As for Leeuwenhoek, he had to explain what the ovaries were and so adapted Aristotle's (and Harvey's) old view that they were equivalent to male nipples: functionless structures that only existed in females because their equivalent (the testes) was crucial in males. And in creatures like birds, lizards, and insects, the egg of the female was nothing but sperm's food.

kind. The word "heredity" with its modern biological meaning was only invented in English in 1863; before that, the Latin *hereditatem* referred mainly to inheritance of status and money.

Maria Sibylla Merian wrote that she would leave such conundrums to "gentlemen scholars." In fact, none of the "gentlemen" of the seventeenth century who worked on generation were genteel eighteenth-century amateurs, much less modern "scientists," a word coined in 1840. They were friends, enemies, artisans, egomaniacs, pioneers to the last. And whether they were "ovists" or "spermists," they all held to a doctrine later called the "doctrine of preformation." Most chuckled at the idea that, with enough magnification, one could find a tiny human crouching inside the head of a sperm, as drawn in 1694 by the microscope maker Nicholaas Hartsoeker. But they did believe that, in some sense, the adult was already there inside the sperm or the egg, whether in miniature or in potential, waiting to unfurl.

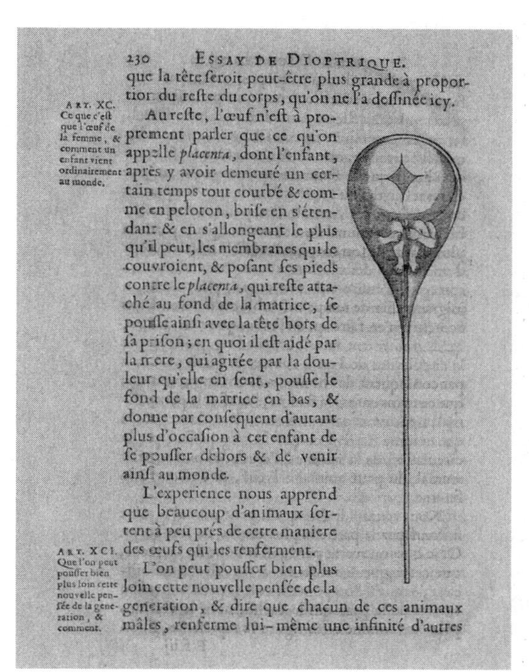

230 ESSAY DE DIOPTRIQUE.

que la tête seroit peut-être plus grande à proportion du reste du corps, qu'on ne l'a dessinée icy.

ART. XC. Ce que c'est que l'œuf de la femme, & comment un enfant vient ordinairement au monde.

Au reste, l'œuf n'est à proprement parler que ce qu'on appelle *placenta*, dont l'enfant, aprés y avoir demeuré un certain temps tout courbé & comme en peloton, brise en s'étendant & en s'allongeant le plus qu'il peut, les membranes qui le couvroient, & posant ses pieds contre le *placenta*, qui reste attaché au fond de la matrice, se pousse ainsi avec la tête hors de sa prison; en quoi il est aidé par la mere, qui agitée par la douleur qu'elle en sent, pousse le fond de la matrice en bas, & donne par consequent d'autant plus d'occasion à cet enfant de se pousser dehors & de venir ainsi au monde.

L'experience nous apprend que beaucoup d'animaux sortent à peu prés de cette maniere des œufs qui les renferment.

ART. XCI. Que l'on peut pousser bien plus loin cette nouvelle pensée de la generation, & comment.

L'on peut pousser bien plus loin cette nouvelle pensée de la generation, & dire que chacun de ces animaux mâles, renferme lui-même une infinité d'autres

The drawing of a homunculus in a sperm head in Nicholaas Hartsoeker's "Essay on Dioptrics" from 1694.

Their advances had been dramatic. In particular, they had established the fixity of species. And they'd learned that animals come from eggs laid by females and inseminated by a male of the same kind. No more dragon teeth sown in the ground turning into a ferocious army of warriors. No more bookworms from winds, or barnacle geese from goose barnacles. "Like breeds like" had become the new wisdom of the early moderns. This was the bedrock on which, in the eighteenth century, Linnaeus could create a new taxonomy, one based strictly on family resemblances.* And yet *how* development worked remained as mysterious as it was to the ancients. The fertilized egg was as featureless as a fog.

One reason for this was because the proper tools needed to look inside a developing egg had not been perfected or invented yet: Microscopes would get much better, and soon early embryologists would begin using dyes. Another was that, despite the coining of the term, no one knew yet that both egg and sperm were actually *cells.* What would be called "cell theory" would turn biology into a new kind of science, more similar to chemistry than to natural history, as it had been practiced in Merian's day. In the coming centuries, a new outlook on life would gradually take hold: Cells are where we all come from. And cell division and death are the forces that sculpt a growing organism.

Over the next one hundred and fifty years, the idea of the fixity of species would gradually give way to a new picture of life. If it had been necessary for the ordering of nature by Linnaeus, it was also necessary for the idea of evolution. For how could species evolve if they were not distinct and separate in the first place? How would they transform from anything if they didn't first breed true? This was a lesson that would have perked the ears of Aristotle: Change could only come about if it rested first on what was stable.

---

* Previous taxonomies had been based on all kinds of categories: what creatures ate, what ate them, their color, whether they had blood or not, where they lived, even the ways they were useful to humans.

Though not as pious as Swammerdam, Goedaert, or Steno, when she died Maria Sibylla Merian still most assuredly believed that God was master of all creation. Every creation, big or small, occupied a fixed rung on the divine ladder. Despite a life devoted to observing caterpillars dramatically morphing into "summer birds," the notion that a species could itself evolve into a different species would have struck her not only as blasphemous, but absurd. Funnily enough, two and a half centuries later, Nabokov, the man inspired by her, would also question evolution by natural selection. And like Merian he'd fall captive to metamorphosis.

In his short story "The Aurelian," Nabokov introduces Paul Pilgram, a "churlish, heavy man, who fed mainly on Erbswurst and boiled potatoes, placidly believing in his newspapers and quite ignorant of the world." Pilgram is a butterfly dealer who has never left his native Berlin, where he owns a little shop "permeated with the pungent odor of a strong disinfectant." He walks with a limp, and his legs seem too thin for his body. His business is failing, his marriage is childless and cold. But Pilgram is a first-class entomologist after whom a rare moth has been named (*Agrotis pilgrami*). And he's a special breed of dreamer, an "Aurelian," whom Nabokov tells us are people who love to find chrysalids, "those jewels of Nature . . . hanging on fences above the dusty nettles of country lanes." Pilgram dreams bigger, of exotic butterflies in Dalmatia and Lapland, even across the oceans, as far away as Tibet. And so he cheats an old widow, selling her late husband's butterfly collection for twenty times what he paid her, and prepares to skip town. He will make his dream a reality, allowing it "to break at last from its crinkly cocoon." Just as he's sneaking out the shop door, perceiving "something almost appalling in the richness of the huge happiness that was leaning towards him like a mountain," he suffers a fatal stroke.

> Yes, Pilgram had gone far, very far. Most probably he visited Granada and Murcia and Albarracin, and then traveled farther still, to Surinam or Taprobane; and one can hardly

> doubt that he saw all the glorious bugs he had longed to see—velvety black butterflies soaring over the jungle, and a tiny moth in Tasmania, and that Chinese "Skipper" said to smell of crushed roses when alive, and the short-clubbed beauty that Mr. Baron had just discovered in Mexico. So, in a certain sense, it is quite irrelevant that some time later, upon wandering into the shop, [his wife] Eleanor saw the checkered suitcase, and then her husband, sprawling on the floor with his back to the counter, among scattered coins, his livid face knocked out of shape by death.

Maria Sibylla Merian died from a stroke while she was working on a new book, leaving behind unfinished watercolors, engraving pens, sketchbooks, and boxes filled with butterflies. Like Pilgram, she is a pilgrim who never reaches her destination. Nabokov helps us see that the journey of the artist-scientist can never be completed, and may exact a moral price.

But like metamorphosis, it can also bring transcendence.

11.

MAYFLY

The mayfly comes to life every May, as its name suggests. In freshwater rivers around the world, it appears, suddenly, on the surface waters, fluttering its wings. Then, it is airborne. It will find a nearby branch and undergo its final molt there. Crawling out of its old skin, it emerges with fully functioning sex organs and prepares for the fight. There is little time left, perhaps hours, no more than a day. Everything has come down to this.

You see, the mayfly appears to humans in May, but it has burrowed in the riverbed for quite some time now. Perhaps two years ago it hatched from a minuscule egg, living an aquatic life mainly unseen by human eyes. It was called a nymph then. In Greek folklore, nymphs were minor female deities, generally regarded as personifications of nature, and often depicted as beautiful maidens. They were not immortal, like goddesses. But they lived much longer than humans.

Underwater, the mayfly nymph eats and grows, eats and grows. Twenty times it will molt, each time looking like a slightly bigger version of its previous self, and a slightly smaller version of the adult. Just before the penultimate molt, it will rise to the surface of the water.

There, its wings appear for the first time. But it has lost its mouthparts and has only underdeveloped sex organs to show for it. The full sexual panoply will appear at the final molt, on a branch or blade of grass on the riverbank, within a matter of hours. But the mouth will not appear again, having completed its underwater task.

Unable to eat, the mayfly lives on its quickly dwindling reserves. In late afternoon, many thousands jump off their branches, taking to the skies above the rivers. They're so dense that they can be detected by Doppler radar, flitting among the riverside trees. Time and again, they will spend their strength darting up into the air, like tiny NASA rockets, then fall downward again like a leaf. This is how they will impress their mates until one is found, serendipitously. Then, in a brief midair tryst, their abdomens will touch, like a kiss.

Now the males fall to the ground, expiring in the thousands on the riverbanks. But the females have one final task. Barely landing on the water, like miniature helicopters out of gas, they release their eggs with one final spurt of energy. As the tiny eggs fall silently to the riverbed, their dying bodies take their final breaths.

Nearly 2,500 years ago, Aristotle asked where creatures come from. For him it was a way to understand cause in an eternal universe, while distinguishing perfect forms from imperfect ones, here on Earth. As Aristotle's ideas came into Christian hands, they continued to delineate man from woman. Answering his question, it transpired, would entail looking carefully at metamorphosis, a surprising key for unlocking a great mystery.

The mayfly has a history going back three hundred million years, almost an eternity. But a mayfly that lives forever is no mayfly at all.

We approach the end of the first trimester. At the systems review, lights off, gel applied, we glue our eyes to the monitor. Tiny hands, a nose, a clear mouth, a beating heart. From head to toes, no more than the size of a kiwi. Our two kids, Shaizee and Abie, are with us, blinking and smiling nervously. I have a clear memory of each of them at this stage, waving specks of life. Now there will be a middle child, and more of them than us. Staring into an Archimedean point in the darkness, I wonder: Are we ready? Any of us?

I've noticed that ever since finding out that she's pregnant, I turn to Yaeli more tenderly. I see her prodigious qualities and inner beauty in plain sight. I know that the root of this is biological, though I'd prefer to think otherwise. Does a mayfly, too, feel the condition of its own existence, the evanescence of the ones it loves?

PART TWO

WHERE ARE WE GOING?

1.

ANNA

The first time he laid eyes on her, at the wedding of his brother Karl and her sister Hermine, she was dressed as an elf. It was September, 1852. At seventeen, Anna Sethe was a year younger than he was, and commandingly beautiful. There was a healthy grace, a natural laugh, an unmistakable air of truth to her. But as he watched her dance, he was filled with melancholy rather than desire. Rather than join in the fun, he sat in a corner by himself.

Ernst Haeckel would one day attempt to solve the seven great riddles of the universe. The result was a book that made him famous, and survived. Gandhi commissioned a translation into Gujarati as a remedy for the wars of religion plaguing his country. Haeckel's other admirers would include the hard-headed Thomas Edison, the dancer and free love proponent Isadora Duncan, the writer D. H. Lawrence, and Sigmund Freud, who made his biology the centerpiece of his psychoanalysis. Lenin would extol Haeckel's work as a great weapon in the class struggle, and his images of deep-sea creatures would shape the sinuous forms of art nouveau. Haeckel would establish a whole new kingdom of life, the Protista, locate heredity in the nucleus, and coin the words *ecology*, *phylum*, and *stem cell*. It was he who set the stage for modern embryology, he who spoke first of the "missing link" between humans and chimpanzees.

A figure of classical proportions, curly-blond and imposing, his calf muscles were as muscular as his beard was stylish, and men and women would fall in love with him with equal ease. Others would despise him, look at him with shame.* During a long and tumultuous life, he would travel the world like his hero Alexander von Humboldt before him, to the Canary Islands, to India, Malaysia, and Ceylon. He would become one of the most widely read authors of the nineteenth century, teaching more people about the theory of evolution than even Charles Darwin himself. A theoretician as brash as he was an observer, he'd describe a law of life that would dominate and divide his community years after his death. Anatomist, marine zoologist, philosopher, and naturalist, he was also a talented artist, and his drawings of embryos would become the most fought over images in the history of modern science. Haeckel sought to depict the world with the precision of a scholar and the passion of a mystic. To arrive at this vision he would need to unite two conflicting worlds—an outer world of rigor and system, to be discovered by science, and an inner world of strange and dark-lit forms, glimpsed by poetry, and art.

But in September 1852 at the wedding, he was still a brooding youth. One who had heeded his father's advice, against his own inclinations, and begun to study medicine. Attending Germany's leading medical school in Würzburg, he'd come across the great biological minds of the age: the master microscopist Albert von Kölliker and the luminary Rudolf Virchow, who taught that all of life was built of cells. Reeling from their lectures, he would climb into bed with tales of his hero Humboldt in the tropics. A romantic, Humboldt was

* Reacting to Haeckel's 1899 book *Die Welträthsel* (*The World Puzzles*), the famed Berlin philosopher Friedrich Paulsen wrote: "I have read this book with burning shame, with shame over the condition of general education and philosophic education of our people. That such a book was possible, that it could be written, printed, bought, read, wondered at, believed in by a people that produced a Kant, a Goethe, a Schopenhauer—that is painfully sad."

as artistically inclined as he was a master measurer, of temperatures and mountains and currents, of birds and trees and soils and tides. Reared at the feet of Germany's greatest philosopher of the previous century, he'd imbibed Immanuel Kant's argument that the universe conformed to some fundamental plan that humans could understand. A grand intelligence was the source of creation, not God in the traditional sense, but a mind existing everywhere and at all times. A mind that needed both science and art to divine.

But Haeckel's august professors were teaching something different. They saw no evidence or need for any master plan. For them, science was science, devoid of any spirit. Nor was art a necessary companion to uncover hidden truths. There were physical principles guiding life, of matter and motion. Any suggestion of an abstract supervising intelligence, or purpose expressed through beauty, should be consigned to the past.

Could there be a middle way? This was the question—a question of crushing, existential import: a spur to scientific research and a beacon of terrible misgiving—that haunted young Haeckel. For this was a moment when biology promised to answer, not only the riddles of life and death, but whether there was meaning in the universe. It overshadowed his every waking hour, even in the midst of a wedding, even in the face of the most beautiful, most divine elf he had ever seen.

Irresistibly drawn to the new science, Haeckel decided to take a summer course in comparative anatomy in Berlin with a man named Johannes Müller, a world-leading physiologist and zoologist, and an insomniac. Unlike Virchow, Müller believed that nature was, indeed, imbued with a vital force. At the same time, Müller was hardheaded: Humanity needed to understand the material, mechanical rules of cause and effect if it was to discern any higher purpose. His course made a big impression. When it was over, in August of 1854, Haeckel traveled with a friend to the archipelago of Helgoland, off the German coast in the North Sea, and was surprised and delighted to come upon his teacher, researching marine creatures with his son. Collecting and dissecting starfish and sea urchins, his heart opened up to the sea. As the four of them fished together in the boat capturing medusae,

Haeckel wondered about the "astonishing generational alternation of these creatures," who turn from tiny, soft, coral-like creatures into jellyfish. Müller looked at him, resigned. "Yes, we are faced with a great riddle!" he said. "We know nothing of the origin of species!"

Back in Würzburg after the summer, Haeckel became Virchow's assistant. From him he was learning that health and disease, growth and decay, all came down to cells. Life itself was the sum of functions of cells making up tissues and organs, each distinctly different in anatomy, but really nothing else. A cold materialism began coursing in his veins, frightening him. He'd been reading the works of a Christian theologian;* perhaps with their help he might be able to reconcile his piety on the one hand and admiration for his teachers on the other by imagining two separate spheres for religion and faith. Contemplating such matters as he sat for his final medical exams, in the spring of 1858, he learned of Müller's untimely death, interpreting it as a suicide.

Shaken to his core, Haeckel returned to Berlin, to write up his dissertation on the tissues of river crabs. It was then he began to keep company with the gorgeous elf from the wedding five years earlier, his sister-in-law's sister, Anna Sethe.

As letter followed letter, afternoon calls and evening strolls, Haeckel slowly began to feel purpose returning into his life. He described Anna to a friend as "a true German child of the forest with blue eyes and blond hair and a lively natural intelligence. . . . She puts no stock in the so-called higher and finer world, for which I hold her even higher since she was brought up in it." This woman of "clear understanding" and "a budding imagination," this "natural person," completely unspoiled and pure, was pulling him back from the precipice of brute materialism. That August he confided in her: "How our souls

---

* Friedrich Schleiermacher (1768–1834) was a reformed theologian and philosopher known for his attempts to reconcile Enlightenment thinking with traditional Protestant Christianity. A modern version of his notion of two separate spheres for science and religion was advocated by Stephen Jay Gould, who used the term *nonoverlapping magisteria*, or NOMA.

have already so closely and strongly grown together. . . . When I press through from this gloomy, hopeless realm of reason to the light of hope and belief—which remains yet a puzzle to me—it will only be through your love, my best, only Anna."

They announced their engagement on September 14, 1858. A few short months later, having spent a miserable year practicing medicine, Haeckel was on a train to Italy, to study the art of the great medieval masters, along with the gems of the Mediterranean Sea. "In Rome I would surely become a pagan," the twenty-five year-old wrote to his love back home. But neither the glories of Baroque saints nor the ruins of classical culture could save him from his state of modern religious doubt. By the time he had arrived in Naples, he'd descended once more into gloom. Gradually, the waters calmed him, and he began once more to find purpose. Studying a daily catch brought to him by a local fisherman, he wrote to Anna: "The fruit of the tree of knowledge is worth the loss of Paradise."

Anna Sethe and Ernst Haeckel shortly after their marriage. Courtesy Ernst-Haeckel-Haus, Jena.

By the start of 1860, Haeckel was back in Berlin. The gorgeous two-volume, hand-drawn monograph he produced on the Radiolaria—single-celled creatures that secrete stunning skeletons of silica, exquisitely minute and diverse—had won him a post as a *Privatdozent* at the University of Jena. He was dazzling students in the lecture hall while taking up serious gymnastics. Gleefully announcing his upgrade to "Archducal-Saxonish-Weimarish-Colburgish-Altenburgish-Miningenish-Extraordinary Professor," with his financial future now secured, he was finally able to offer Anna the prospect of marriage. "The more I attain inner calm and clarity here," he wrote to Anna,

> through energetic external activity, as well as through lively mental exercise, and the more the peace of nature is drawn into my soul, the clearer it becomes to me what a great, inestimable, enviable happiness has bloomed in me during these last years in which I have possessed the loveliest, purest maiden soul and the most noble, most beautiful friendship, and these continue to mature into ever more blossoms and happy fruit. Love and friendship! How happy they make me. I had earlier chosen science alone, but they promise me everything that science cannot give.

When they were married on August 18, 1862, Haeckel wrote to his parents that with each day he was more in love with Anna, though "cannot believe that any more love is possible." Then, less than a year and a half later, Anna suffered from a severe inflammation around her lungs. Within days, this became a pain in her stomach, which spread to her liver. Losing consciousness late in the morning of February 16, 1864, she died that afternoon, perhaps of appendicitis, at just twenty-eight years of age.

The catastrophe was too swift, too deep, too dark to be borne. Haeckel fell unconscious and spent days in bed, in and out of delirium. He spoke of suicide, and wrote to a friend: "I am dead on the inside already and dead for everything. Life, nature, science have no appeal to me. How slowly the hours pass."

Distraught, his parents sent him to the Mediterranean coast of Nice; he had been there once before in happier times, and they hoped it might lift his spirits. Alone by the sea, Haeckel shuffled down a pebbled beach, in agony. As recounted by his biographer, Robert J. Richards, he happened to glance indolently down at the surface of a tidal pool when his puffy eyes fell upon a species of medusae he had never seen. He froze. In the glint of the sun bouncing off the ripples of the pool, the long tentacles hanging over the rim suddenly looked to him like blond locks of hair. It was her. He was sure of it. It was Anna.

Years later, remarried and world famous, Haeckel would recall that moment in 1864 in the Bay of Villafranca, when he encountered "the most charming and beautiful of medusae":

> I name this species, the princess of the Eucopiden, as a memorial to my unforgettable true wife, Anna Sethe. If I have succeeded, during my earthly pilgrimage in accomplishing something for natural science and humanity, I owe the greatest part to the ennobling influence of this gifted wife, who was torn from me through sudden death.

Haeckel called the medusa *Mitrocoma annae*. Soon he too would return from the dead—to the delight and dismay of many—and found his own religion.

2.

IMMORTAL JELLYFISH

There's a creature in the ocean that lives its life backward. Or, to be more precise, it grows forward and backward, old and then young and then old again, in alternation. It begins life innocently enough, as an egg, just like everyone. But unlike the jellyfish we avoid at the beach, this one spends only a short period of its life as medusae. After it's fertilized, it grows into a minuscule ball of cells called a planula, and settles on the seabed, giving rise to a colony of asexual polyps. Connected by horizontal stemlike structures called stolons, affixed to the seabed, the polyps are all perfect genetic clones. Swaying gently in the current, their tentacles pointing upward, they look like sprigs of dill with tiny buds and an attitude. Then something unexpected happens: Rather than sprouting flowers, the buds swell and begin to sprout upside-down, disc-shaped creatures in the tens and hundreds: immature medusae just 1 millimeter long. Drifting in the waters, they feed on plankton, then flip over. Depending on the water's temperature, within three to four weeks these undulating domes will become sexually mature adults. Translucent save for a bright scarlet bell at the core of those that swim in cold waters, they sprout eighty to ninety haunting, dangling tentacles that will help them hunt for food. They're fully grown now, these otherworldly creatures. Just 3 millimeters thick, between a nickel and a dime.

This is when strange things start to happen. Because unlike other hydrozoans that expire once they've released egg and sperm to form a new generation, these creatures never really die. Instead they keep on living, and living, and living some more. They do this by a process called *transdifferentiation*, which is the conversion of one cell type into another. And so, when threatened by physical injury, or stress or starvation, the mature medusa reabsorbs its tentacles, shrinks, and begins to sink downward. Sacrificing its ability to move, and its sexual organs, it flips upside down, folds into itself, forms an outer shell, and latches to the ground as a polyp. Spectacularly, the adult is a baby once again.

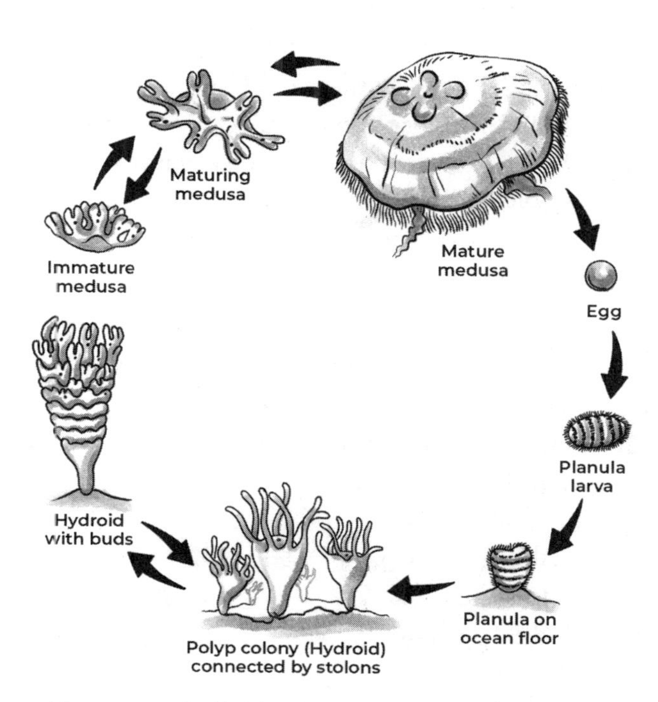

The immortal jellyfish life cycle. Credit: Inna Gertsberg.

Try to wrap your head around it. What has just happened is equivalent to a butterfly changing back into a caterpillar after having emerged fully formed from its chrysalis. More precisely, it's like a chicken reverting back to an egg, which then gives birth to the same chicken. That's because, as the buds swell, identical clones of the original jellyfish will once again be released by the polyp into the water column. Again these immature medusae will flip, growing into disc-like domes. And when these tentacle-toting adults feel threatened, they too will flip, fold, form an outer shell and turn into polyps. The cycle will be repeated, over and over, staving off the creature's demise. A single jellyfish may die in the mouth of a fish, or from a virus. A polyp may be nibbled by its nemesis, the sea slug. But the genome is here for good.

The immortal jellyfish, as it is often called, was discovered in 1988 by a German marine biology student snorkeling off the coast of Rapallo in the Italian Riviera, just a three-hour drive up the coast of the Ligurian Sea from where Haeckel had gone to recover from Anna's death, in Nice. Precisely a century before its discovery, and just a few decades after Haeckel's fateful visit, this was the place where Friedrich Nietzsche would conceive one of his most famous books. "Everything goes, everything comes back, eternally rolls the wheel of being," he wrote in *Thus Spoke Zarathustra*, with no idea that he was describing the local fauna.

The immortal jellyfish never dies of old age. Instead, like Benjamin Button, it appears to age in reverse. The species is far from finicky: Able to live in many seas, it's spreading from the Western Mediterranean to the shores of Florida, Panama, Israel, and Japan, often hitchhiking on cargo ships that use seawater as ballast. It is rapidly growing in numbers. Few have taken notice, but at this rate what the Notre Dame biologist Maria Pia Miglietta calls the "silent invasion" may seriously transform our planet. As Nathaniel Rich commented over a decade ago in *The New York Times Magazine*, "It is possible to imagine a distant future in which most other species of life are extinct but the ocean will consist overwhelmingly of immortal jellyfish, a great gelatin consciousness everlasting."

In light of this scary prospect, it would seem a good idea to know something about this unusual creature, whose scientific name is *Turritopsis dohrnii*. But, its deathlessness notwithstanding, it is actually extremely difficult to raise in the lab. Luckily, an extrovert named Shin Kubota from the sleepy beach town of Shirahama, four hours south of Kyoto, has taken on the project. A karaoke enthusiast who is known to sing during his scientific talks, Professor Kubota has devoted his life to the jellyfish. He and a small group of hydrozoan experts around the world believe these creatures have struck an evolutionary bargain: Forsaking a brain, a heart, and eyes, effectively eating out of their anus, they have gained immortality in return. Feeding his tiny jellyfish daily with even tinier food he cuts up like a doting father, he has been heard shouting at individual medusae, "Eat by yourself!" "You are not a baby!" or "You, rejuvenate!" If only we could learn more about *Turritopsis* development, he believes, we might be able to overcome cancer, senescence, maybe even death itself.

With little funding, Kubota has been able to learn a lot that's worth knowing. How to feed his charges dried brine shrimp eggs harvested from the Great Salt Lake in Utah, in little petri dishes placed in refrigerators, keeping them happy and reproducing. How to inhibit rejuvenation by starving the little creatures, or placing them in water colder than 72 degrees Fahrenheit. Kubota has also found a way to induce the immortal jellyfish to turn backward, incessantly stabbing the medusa's bell with two fine metal picks. When mutilated in this way, *Turritopsis* will lie limply on its side, its bell deflated, its tentacles ceasing to ripple in the waters. But then the miracle happens. Within a few days, it reabsorbs its tentacles into its body. Stolons begin to shoot and attach themselves to the petri dish. Before long, the assaulted adult is a polyp again, having gone through the steps of transdifferentiation. It's a primitive method, and gory, but it works.

Kubota and his colleagues have published a full genome analysis of the immortal jellyfish, and in other labs far away from the sleepy town of Shirahama, scientists have begun to unlock the genetic secrets that give hydrozoans everlasting life. As reported by Rich, one researcher, the molecular paleobiologist Kevin Peterson at Dartmouth College, believes small RNA molecules, called micro or miRNAs, play an important role, acting as on/off switches that control whether a primitive, undifferentiated cell will turn into a specific mature cell—part of a tentacle, say, or a foot. MicroRNAs are known to play a crucial role in humans: When they change spontaneously they can upset the usual division regimes of cells, making them go berserk. Studying how miRNAs control differentiation is already playing a role in fighting human cancer. Peterson says we shouldn't be surprised: At a genetic level "we look like a damn jellyfish."

More recently, the Israeli geneticist Erez Levanon has described a different mechanism called A-to-I RNA editing, which not only alters the cellular fate of RNA molecules but also changes their sequence relative to the genome. That means that with tweaks to their regular RNA, creatures might be able to produce an enormous amount of variation without at all changing their DNA. One could imagine how this might be useful for strengthening a creature's immune system, but it might also have another function, altogether: Producing a large amount of dramatic variation quickly is precisely what's necessary when a creature changes from one form into another during metamorphosis.

When it comes to defeating mortality, though, rather than *Turritopsis*, another hydrozoan species may lead the way. Named for the serpentine water monster of Greek mythology whose many heads, when chopped off, would grow back many times over, the *Hydra* lives in freshwater, looks like a polyp, and never produces medusae. Instead, it has opted for a simpler lifestyle. Why this is so has been the passion of a researcher named Daniel Martínez, who set out in his doctorate to prove that *Hydra* are not immortal. Martínez ended up changing his mind. What he discovered was that *Hydra* have not one but three stem-cell populations, continually renewing themselves at the core of their bodies. The differentiated cells of the tentacles and the foot, on the other hand, are constantly brushed off and replaced with new cells migrating from the body column. Humans have long assumed that mortality rises with age, but that might not always be true. "I do believe that an individual *Hydra* can live forever under the right circumstances," Martínez says.

When Anna died, Haeckel wrote that her death "destroyed with one blow" the remains of his dualistic worldview. Haeckel's biographer believes that after the tragedy, there could be no escape for him from brute materialism, selection, and chance. This worldview he found in the teachings of Charles Darwin, whose own loss of faith has been attributed to the tragic death of his eleven-year-old daughter, Annie. Now Haeckel's energies became focused on just one task—bringing the light of evolution to the world. His accomplices would be the sea creatures: armored Radiolaria and enigmatic Cnidaria, the marine phylum that includes the immortal *Hydra* and *Turritopsis*. It was in these hauntingly beautiful creatures that Anna continued to live on in Haeckel's imagination, forever resisting death.

Haeckel would go on to become the nineteenth century's most important Darwinian, second only to Darwin. But to see how this happened, we have to rewind in time, like a *Turritopsis*. There's a figure whom we haven't really met, but who already casts a shadow over this story. Touting the dual nature of a universe suffused with mind and matter, he represented everything that Haeckel spent his life struggling over. He was heading toward a brutish worldview, but the ghost of his greatest hero continued tugging him back.

At the root of everything for Haeckel was Johann Wolfgang von Goethe.

3.

GOETHE

Goethe is obsessed. He wants to know, as he writes to a friend, "the truth about the *how* of the organism." How it is born, how it grows, how it feels to be what it is. *The Metamorphosis of Plants* was his first and major attempt at it. Despite nature's diversity, he is certain now, in 1790, that there must be a unity, an ur-plan to bring all forms together, a logic behind their making. But nature can only be penetrated if the scientist becomes one with his subject: He will need to gain insight from within.

Goethe, in fact, is more than a scientist. Or rather, simultaneously both more and less. He has started out as a lawyer, like his father, establishing a small legal practice in Frankfurt. But his heart is elsewhere, away from dull formalisms and precedents. As a child he was infatuated with puppet shows, with the five books of Moses, Virgil's *Aeneid* and Ovid's *Metamorphoses*. He loves languages, and theater, history and the arts. Discovering Shakespeare, a self-reported transformative event, he begins to write poems himself, even a drama. But what will save him from a life in law is a very short novel.

The Sorrows of Young Werther, a tragic tale of unrequited love, immediately catapults Goethe into the literary stratosphere (Napoleon will famously take a copy with him when he sails off to conquer Egypt). Establishing German for the first time as a European literary language, it becomes a cult book for a generation. The following year, 1775, the grand duke of Saxe-Weimar-Eisenach invites Goethe to his court. Goethe is twenty-six, the duke just eighteen, but Karl August has read *Werther* and, like the rest of Germany, is charmed. Appointed to his Privy Council, Goethe assumes a rather astonishing list of official duties at court: As head of the Duchy's War Commission, he recruits mercenaries to fight against the American revolutionaries; as head of the Mines and Highways Commissions, he oversees the digging of silver and building of roads; he even becomes chancellor of the Exchequer. He helps plan Weimar's botanical park, and implements reforms in its university at Jena. In 1782 he is ennobled, becoming von Goethe. He is now the duke's closest confidant and friend.

The future holds great things for him. He will write the magnificent novel *Wilhelm Meister's Apprenticeship*, about a bourgeois businessman's empty life and journey to self-realization (the romantic critic Friedrich Schlegel will judge it equal in importance to Europe as the French Revolution). He will also write his dark poetic masterpiece, *Faust: A Tragedy*, the tale of God's favorite human being on earth, the scholar Heinrich Faust, selling his soul to the Devil in exchange for earthly pleasures and unlimited knowledge. Goethe, like Faust and the alchemists of old, aspires to encompass the whole universe and overcome all its contradictions. His breast bursts with passion, but he knows that life is full of sorrow. "While Man's desires and aspirations stir," he'll have his protagonist lament, "He cannot choose but err."

But Goethe feels trapped. Fleeing the limelight of his transcontinental celebrity, he decides to travel to Italy under an assumed name. It is 1786, and the next twenty months will forever change him. Alongside the discovery of the treasures of Roman and Greek antiquity, alongside the paintings and sculptures and architecture later adoringly described in his *Italian Journey*, Goethe is immensely struck by the beauty of the outdoors. For readers of Ralph Waldo Emerson, which is to say, most enlightened Americans of the second half of the nineteenth century, Goethe will be remembered as one of the six "representative men" of humanity—"the Writer," alongside Plato, "the Philosopher"; Swedenborg, "the Mystic"; Montaigne, "the Skeptic"; Shakespeare, "the Poet"; and Napoleon, "the Man of the World."* He is revered even today as the founding figure of Romanticism, straddling the classical and modern worlds. But for the moment, Goethe puts aside his voracious appetite for human affairs. As he makes his way from Venice to Sicily, Verona to Bologna, Palermo to Padua; from the Adriatic to the Apennines to the Straights of Messina and back again to Rome, he is now desperately obsessed with flowers.

Why flowers? Before Italy, Goethe had noticed the fundamental similarity of structure among different organisms: Just as humans have femur bones, so, in slightly different proportions, do lions, hippopotami, and dogs. Anatomists of his age had observed that vertebrates, including monkeys, all had a bone in their jaws called an intermaxilla—all, that is, except for humans. Prevailing wisdom took this to be a distinguishing mark, a dividing line that separated us from the beasts, but Goethe wasn't convinced. In the spirit of the teachings of towering Kant, there had to be a unity, a graspable plan underlying all variations. Locating the intermaxillary bone tucked away in the upper jaw of humans with cleft palates, in whom it was more conspicuous, he was vindicated: Clearly this was a sign of our connection to all life.

Goethe's Italian journal is filled with observations on the geology of the southern regions. He writes of minerals and rocks taken from riverbeds and mountaintops. He even climbs, treacherously, to the summit of Mount Vesuvius, cataloging the various lava flows while doing his best not to turn into cinder. But Goethe's true passion lays in what he calls the *Urpflanze*. Just like in animals, in the plant world there had to be an underlying unity. "Among this multitude," he writes of Italy's bursting flora, "might I not discover the Primal

* In Ralph Waldo Emerson's seven lectures, "Representative Men," published in 1850.

Plant? There certainly must be one. Otherwise, how could I recognize that this or that form was a plant if all were not built upon the same basic model?" But what is the basic model, the *Urpflanze* that gave rise to untold varieties? This is what he wants to know.

Upon return from the land of forms to "formless" Germany, exchanging a bright sky for a dark one, Goethe writes that his spirit "sought to escape injury through intense rebellion." In Rome he had met the painter Johann Tischbein, who rendered him reclining among ancient ruins, contemplative and serene. It is by far the most famous portrait in Germany, recognizable to schoolchildren. But Goethe returns from his Italian journey unnerved. He had glimpsed an insight in the leaves of a palm tree at the botanical gardens in Padua that had made his heart thump, and had experienced at Palermo an intense inner feeling of connection to the Proteus* of all plants. "The archetypal plant as I see it will be the most wonderful creation in the whole world, and nature herself will envy me for it," he'd written from Italy to the philosopher Johann Gottfried Herder back home, sounding more like a twenty-first-century genetic engineer than an eighteenth-century poet. "With this model and the key to it, one will be able to invent plants . . . which, even if they do not actually exist, nevertheless might exist and which are not merely picturesque or poetic visions and illusions, but have inner truth and logic." Now he sets out to put it all on paper: his philosophy of change married to his philosophy of science.

There is an inner unity to life, and a spirit pushing it forward. Forces at work that bring about all the diverse forms, compelling them to blossom and bloom. The Englishman Isaac Newton had placed calculation above all else. The Swede Linnaeus had imposed an artificial taxonomy on nature, grouping animals and plants according to things he could measure: number, size, proportion, form. But nature would not be so anatomized nor entirely quantified; she was

* Envisioning the act of creative expression welling up from the depth of the ocean, Romantic poets often invoked Proteus, a Greek god of the sea capable of assuming an infinite variety of forms.

a dynamic process, not a final product, as much beholden to history as to transcendental purpose. Crucially, what was true for hyacinths was also true for humans: By becoming one with flowers, Goethe hopes to find his inner way of being, he writes, "to reveal as it were myself."

He will accomplish this by breaking out of "the grim torture chamber of empiricism, mechanism and dogmatism," to "live into" the natural world. The Age of Reason had produced a static view of human nature, a mechanical understanding of the universe, a linear model of the advancement of knowledge. This might have been good for Copernicus, Galileo, and Newton, studying *natura naturata*, or the inert world of objects. But when it came to *natura naturans*, vital nature in constant flux and flow and transformation, a strictly physical-mechanical philosophy could only impoverish man's understanding. Descartes, turning inward to the mind, likewise offered a philosophy that was sterile, for how could one learn anything without first looking about? Neither empiricism nor rationalism could bridge the gap between the objective world and the subjective mind learning it. Kant had already defined the problem: What we perceive is only a mental representation of the world, not the world itself. So are we eternally damned never to know reality?

To penetrate the real, living, shimmering world, Goethe will need to embark on an "adventure of reason," one in which "the labor of experimentation," he writes, is intimately connected to human values and development. Cartesian-Newtonian science separated observer from observed, but Goethe comes to see that far from a requisite, the separation is a barrier. In an interactive experience not only the object of observation changes but also the observing subject. A true science of vital nature has to be vital itself, dynamic and ever-changing; it is meant as much to bring about a metamorphosis in the scientist as to inform his view of the world.

One day they'll call it Goethean, or Romantic Science, but he is toiling now on its method. Using what he renders an "exact sensorial imagination," he calls upon his inward artistic side to sense the fluid processes of growth and transformation. Rather than through cold external analysis, he is guided by reciprocity, wonderment, and gratitude. He respects but suspects quantification, complementing it with creative thinking to penetrate a reality beyond the senses. That's how he discovers in himself nature's multitudinous reflection. And begins to gain a new way of knowing the world.

From his work for the grand duke planning the botanical park of Weimar, Goethe is familiar with plants whose stamens can turn into petals, creating flowers with many rings of petals but no sexual parts. An example is roses, which humans had cultivated to create attractive garden varieties. But there are other flowers whose petals turn into stamens, leaving their sexual parts unprotected. Like looking at humans with cleft palates to discover the intermaxillary bone, it takes looking at these "abnormal" cases to appreciate the way things are when all goes well. If stamens and petals are interchangeable, flowers and leaves are, in fact, the same thing. To understand how this could be, all one needs to do is look at a plant's development: It begins in the ground, where out of a seed come two cotyledons, or tiny embryonic leaves. The cotyledons are separated, but their ultimate fortune is to be united. For in the course of the development of the axis, the shoot will begin to climb upward toward its climax. Lateral appendages will appear, periodically, as if by preordained fiat. And just like a succession of limbs in a millipede paired along an axis, or the sequence of vertebrae in a human leading up to a head, serially repeating structures will begin to produce slight transformations. All the while, the plant will be purifying its sap as it climbs higher and higher. "Everything material, more paltry and vulgar is gradually left behind," Goethe writes in one of the one hundred twenty-three short, aphoristic paragraphs that make up *The Metamorphosis of Plants*, "while that which is higher, more spiritual and superior is allowed to display itself with greater freedom."

It happens like an accelerating drumroll, announcing an impending drama. The first appendages up the axis are foliage leaves, spaced incrementally above each other. But after a few series of these, the rhythm heightens, and suddenly something new appears. It's a slightly transformed leaf, growing on the stem, one we call a sepal. And within an inner whorl directly above it, as if impatient to appear, the corolla of colorful petals now blossoms into existence. Next are the stamens and carpels, all clustered within. The female and male cotyledons have finally been united. The plant is at its zenith, ready to bring about new life; producing a seed, it has come as close as it can ever come to transcending time itself. Whether it happens to be a lily or gardenia, a periwinkle or rose, all its sundry parts have emanated from the very same motif. There is a rhythm and a logic to everything, a beginning and an end with a grand apotheosis. The plant has sprouted, expanded, and finally contracted. Graduating from imperfect potential to perfect realization, it has followed its spiritual path, just as a human should.

"Everything is leaf," Goethe writes, undoubtedly contemplating his own origin. Life begins with an abstract generating power, an ur-plan, and then forges untold variations. "There is in nature an eternal life, becoming, and movement," he writes, "she alters herself eternally and is never still. She has no conception of stasis and can only curse it."

Goethe is fascinated by the similarities that transcend taxonomy—by how much all life-forms have in common. Just as leaves are for plants, the spines of vertebrate animals are elemental structures; skeletons and skulls of lizards, wolves, and humans are merely different versions of the elemental unit metamorphosed. He shares the inevitable conclusion in private with his good friend Alexander von Humboldt. In public, almost three decades before Darwin's *Origin of Species*, he writes, in *Story of My Botanical Studies*:

> The ever-changing display of plant forms, which I have followed for so many years, awakens increasingly within me the notion: The plant forms which surround us were not all created at some given point in time and then locked into the

given form they have been given . . . a felicitous mobility and plasticity allows them to grow and adapt themselves to many different conditions in many different places.

He dies just a short year later, on the brink of mysteries untold. "*Mehr Licht!*" he is said to have asked just before his heart failed.

More light. Please, just a little more light.

4.

PROGRESS

Historians claim that the idea of progress was born in the time between Christopher Columbus and Isaac Newton.* Hierarchy, on the other hand, is as old as the wind. In ancient Greece, for example, Aristotle held that there was a pecking order in nature: sensible and motile animals existing above sensible but stationary plants; animals that give birth to live young above those that lay eggs above those that produce larvae; warm-blooded creatures above the cold-blooded, themselves above the invertebrates, altogether "bloodless."

Aristotelian zoology was a secular zoology, but it was taken up in the Middle Ages by philosophers who rechristened it the "Great Chain of Being." Climbing up a ladder from the basest rocks and minerals through the plants and fish and insects and lizards and birds and mammals to humans to kings and queens, the rungs brought one closer and closer to the heavens, where the angels and God resided. But this continuum was a fixed order, unmovable and preordained. Alchemists might theorize that all matter was made

* Joel Mokyr, *A Culture of Growth: The Origins of the Modern Economy* (Princeton, 2016).

of the same elemental stuff, such that a base metal, treated the right way, might be transformed into precious gold. But no one questioned the rigid structure of the cosmos, or that it was all ordained by God.

Nowhere did this static attitude manifest more clearly than in humankind's collective dreaming. Think of the word *utopia*, Greek for "no place," or "nowhere." Or of the Golden Ages of the ancient world, of the Garden of Eden, and arcadias, the Isles of the Blest. These were places where humans and animals are united, birth is painless, and there exists no death. But such places are gifted, they are never achieved. They take place at the End of Days, or back at the Origin, not anytime or anywhere in anyone's lifetime, or even death. History is excluded as a medium for collective striving. Politics can be better or worse, and the moral sphere, too. But fate is not something humans control.

Fate, however, may be repeated. The Greek historian Polybius asserted in his *Histories* that, like the life cycle of birth, childhood, youth, maturity, old age, and death characteristic of individuals, there was a rhythm to human societies, from monarchy to aristocracy to democracy and back again. Due to the special weaknesses of each form of government, it trapped humans in an endless cycle of tyranny, oligarchy, and rule of the mob. For some it might sound better than a preordained plunge from a Golden Age, or Fall from the Garden of Eden—though many held out for redemption at the End of Days. Whatever the sensibility, misery and folly were givens. Little room was left for humanity to bring about true change.*

Science and technology began to convince people that things could be otherwise, that if we put our minds to it, we could better our station, here, in this world. After all, Ptolemy had no telescope, Pliny no microscope, and Archimedes no barometer. To begin to take the future seriously, we would therefore need to do one better than the Renaissance, which had set itself the limited goal of recapturing

---

* This doctrine has never truly left us. Consider Machiavelli, Vico, Spengler, and Toynbee. And before Polybius, consider Plato.

ancient wisdoms from the past. We would need to climb over Plato, for whom this world offered only feeble imitations, like shadows on a cave wall. We would need men, in other words, like Christopher Columbus and Isaac Newton. Men who went out to the world to unlock its secrets, using their God-given intelligence and the scientific method. Then, as Francis Bacon promised in his 1626 book *New Atlantis*, we might go about forging a brave new world.

All this, inevitably, would lead to the replacement of theology by history. Theology treated time as a series of events wrought upon us from above; God's will, not the will of humans, was what determined the ebbs and flows of fate. History would conceive of time as a chain of occurrences brought about by humans. It would be studied so that such humans, endowed with free will, could actually learn from it. The replacement of theology by history would take time and would never be entirely completed, but by the eighteenth century the process had already gotten underway. What would be the road to progress? Gradually, in the West, three traditions presented themselves. The French called for the adoption of the ideals of their Revolution—*liberté, egalité, fraternité*, guided by reason. Swept by industry and invention, the British touted the "Invisible Hand" of free markets, which, as Adam Smith put it, would by serendipitous logic "advance the interest of the society, and afford means to the multiplication of species." The Germans, for their part, offered a nature philosophy that took progress to be the result of purposeful laws of the universe. Out of the tension of opposites, the philosopher Georg Wilhelm Friedrich Hegel argued, would come a synthesis of something new, something higher, more evolved.

The French were revolutionaries, the English pragmatic, the Germans idealists.* Between them, the Judeo-Christian view of history as static until deliverance, and the Greco-Roman view of history as cyclically preordained, had given way to the modern ideal of progress. The ancients had inquired after man's place in the cosmos; now moderns, as Bacon had instructed, could "conquer nature in action." A change had come about in the understanding of change itself.

---

* This is a simplification.

Perhaps it isn't all that surprising, then, that there began to appear chinks in the "Great Chain of Being"; Aristotle's Christianized *scala naturae* now seeming to some less sacrosanct than previously supposed. A man as sophisticated as Thomas Jefferson could still state unequivocally in 1770 that not a single species had been lost since Creation. But the world's leading naturalist around the mid-eighteenth century, Georges-Louis Leclerc, Comte de Buffon, assailed Linnaeus's definition of species as too static.* What most people thought of species were actually varieties modified from an original form; Buffon was adamant that they could "improve" or "degenerate," depending on their environment. Before the century was out, Leclerc's countryman, the zoologist Georges Cuvier, published findings comparing living elephants to fossils of mastodons, ending the argument about whether or not creatures could go extinct. They could, and they had, and in the future other creatures would, too. The conservative Cuvier didn't believe in evolution, but in cycles of creation and destruction. Catastrophism, as he called it, meant that whatever the Bible said, there had been many more than just one flood. Over in Scotland, the geologist James Hutton agreed: Earth was far more ancient than Adam and Eve.**

The idea of a world characterized by constant change was not, in itself, a new invention. The Chinese had believed for centuries that humans, nature, and the heavens were in a state of constant transformation, a concept known as *Tao*. The Greek world produced Empedocles, and the Roman world Lucretius, both of whom explained the origin and undeniable development of the universe, with no recourse either to purpose or the supernatural. Even the medieval

---

* Linnaeus did believe that hybrid plants could produce new species, but Joseph Kölreuter showed in the 1760s that hybrids don't breed true and therefore cannot produce new species. It was the true breeding species that Linnaeus thought were forever stable, but he was wrong. On hybrids producing new species he was actually right, if they're able to become isolated from their parent populations.

** The Christian theologian Augustine of Hippo had encouraged believers to read the Bible allegorically ever since the fourth century: One day in the eyes of God, he argued, might be equal to an eternity.

Muslim polymath Ibn Khaldun, in his influential book the *Muqaddimah*, asserted cryptically that "humans developed from the world of the monkeys" by way of a process that made species more plentiful. The historian Rebecca Stott has beautifully documented these and many other very old departures from the notion that life was set in stone at the moment of Creation. The idea of evolution evolved over long stretches of time, just like plants and animals. Its development was never linear. It was mostly religious, though sometimes secular. And nothing about its history was preordained.

Still, in the first half of the nineteenth century, people came to read the book of nature in a new way, informed by the idea of progress. Étienne Geoffroy Saint-Hilaire had accompanied Napoleon's campaign to Egypt as an official zoologist for the French forces. He was a morphologist, not an evolutionist, and he argued like Goethe that related architectures—fish, birds, reptiles, and mammals—had a Unity of Type. He observed that unrelated groups, like vertebrates and mollusks, also shared the relative position of certain structures, and that all plants and animals, when acted upon directly by their environments, could change to a certain degree. Geoffroy's countryman and older colleague, Jean-Baptiste Lamarck, had gone much further. For him life was constantly climbing on preordained trajectories, from simple beginnings to lofty heights. Minus the British bravado, he sounded a lot like the eighteenth-century doctor, poet, and polymath Erasmus Darwin, who, in his *The Temple of Nature*, published posthumously in 1803, wrote this verse:

> *ORGANIC LIFE beneath the shoreless waves*
> *Was born and nurs'd in Ocean's pearly caves;*
> *First, forms minute, unseen by spheric glass,*
> *Move on the mud, or pierce the watery mass;*
> *These, as successive generations bloom,*
> *New powers acquire, and larger limbs assume;*
> *Whence countless groups of vegetation spring,*
> *And breathing realms of fin, and feet, and wing.*
> *Thus the tall Oak, the giant of the wood,*

Which bears Britannia's thunders on the flood;
The Whale, unmeasured monster of the main,
The lordly Lion, monarch of the plain,
The Eagle soaring in the realms of air,
Whose eye undazzled drinks the solar glare,
Imperious man, who rules the bestial crowd,
Of language, reason, and reflection proud,
With brow erect, who scorns this earthy sod,
And styles himself the image of his God;
Arose from rudiments of form and sens
An embryon point, or microscopic ens!

Erasmus Darwin wasn't clear about how life began; Lamarck, on the other hand, touted spontaneous generation. But both believed in the inheritance of acquired traits. And while Geoffroy thought that creatures were changed by their environments, Lamarck once more went further: Individual organisms could improve their fortunes by changing their habits, by trial and error, or even volition. If these habits were passed on to the next generation, over time the species would progress. Though this idea would come to carry his name—Lamarckism—and would spark lasting controversy, the mechanism of species change was less important to Lamarck than the direction in which all such changes tended. The inexorable march forward from simple to complex was the main "life force," independent of circumstances. And the pinnacle of that process was Man himself.

Meanwhile, over in England, the geologist Charles Lyell began to argue that Cuvier's catastrophism was short-sighted and melodramatic. The Earth changed gradually over unimaginable tracts of time, by the same slow forces of weathering and erosion we observe today—a theory he called *uniformitarianism*. Walking the halls of the Royal Society beside him was a man named Richard Owen, the nation's leading anatomist. Sir Richard was famous for coining the word "dinosaurs," and for his cockiness. Like Lyell, he was a conservative, believing that each natural species was a perfect idea hatched in the mind of God. But Sir Richard was impressed by Geoffroy, and coined another word—"homology." The wing of the bat, the paddle of the porpoise, the hand of the monkey, and of man, all had an underlying unity. A sign, certainly, of a celestial plan.

Up in Edinburgh, a doctor turned marine zoologist demurred. There was certainly an underlying unity, but it had nothing whatsoever to do with God. Robert Grant was a materialist and an atheist, a radical who championed French School *transformationism* alongside political reform. When a young man named Charles Darwin arrived in Edinburgh at his father's behest to study medicine, Grant and Darwin found each other and began frequenting the estuary at the Firth of Forth. There, in the outdoors, Grant taught Darwin to dissect sea creatures under the microscope, to look carefully at every tiny feature, note every minute variation. Quickly the protégé discovered that dissecting mollusks and crustaceans appealed to him much more than human cadavers. A year of medical school would suffice; it was down to Cambridge now to read theology, with the hopes of retiring to a calling familiar to his times: a vicar-naturalist in a quiet parish. Famously, Darwin's plans were punctuated by a historic four-and-a-half year journey to the southern seas. It would change everything for him, even more for the world around him. And Grant's materialism, Lyell's gradualism, Sir Richard's homology—all would play their part.

When Darwin did return from his voyage on the *Beagle* in October 1836, he set out to crack "the mystery of mysteries." He came back wanting to know where species had come from and how they changed. These were hard questions. If clues were available, it mattered little whether they originated in reactionary or radical quarters—as long as they helped provide a solution. Settling back in London, and then in the village Downe, Darwin began corresponding with dog and pigeon breeders, and performing experiments of his own. He read Malthus (another reactionary), as well as Lamarck, Sir Richard, Geoffroy, Goethe, and his own grandfather, Erasmus. He read the *Vestiges of the Natural History of Creation*, too, a haphazardly argued, anonymously published book from 1844 about how the evolution of the solar system and life on Earth had all been a preordained plan leading up to man. Here was a law of progress, the author claimed, woven into the very bones of the universe.* Slammed by materialists for suggesting a higher force governing fate, *Vestiges* was also condemned by conservatives, who didn't fancy man depicted as a mere pinnacle of an animal process. Such reactions made Darwin tremble. Conflict-averse, and happily married to a religious woman, he fully comprehended the stakes.

When it came to the question of progress, Darwin was of two minds. On the one hand, natural selection would cull the less fit while tweaking form and behavior slowly to render the rest better adapted for survival. On the other hand, species lived in local environments, each one unique, so what could it mean to speak of a necessary climb "upward"? Crabs move sideways. Were they in any sense "lower" than lizards who run straight across the water, or snakes who slither in the form of an *S*? It did seem from the fossil record that creatures were growing increasingly more complex, but hadn't moles gone underground, losing their eyes, and didn't the lowly bee seem hugely successful, if only to judge by its numbers? Darwin was torn. "The simplest cannot help . . . becoming more complicated," he wrote in one place, "and if we look to first origin there must be progress." In another he stated emphatically: "It is absurd to talk of one animal being higher than another—we consider those where the cerebral structures intellectual faculties most developed, as highest. A bee doubtless would where the instincts were."

The static Great Chain of Being was being laid to rest, replaced by a dynamic Tree of Life. But in which direction was the tree growing? And where were humans to be placed? Once otherworldly, cyclical, or devoid of meaning altogether, now change became change for the better.

Or so, at least, some dared to think.

---

* Herbert Spencer (1820–1903) was the most famous public intellectual of the Victorian age writing in this vein, often referred to, alongside Auguste Comte and Henri de Saint-Simon, as one of the nineteenth century's "prophets of progress." The anonymous author of *Vestiges*, on the other hand, turned out to be the Scotsman Robert Chambers, influential in political and scientific circles, mainly as a publisher and geologist.

5.

THE EL DORADO OF ZOOLOGY

Ernst Haeckel was destined to be a believer.

Some looked at the heavens and the earth and found no God there, just sterile physics. Others saw God shining gloriously from within Newton and Euclid's laws. But Haeckel knew that what men called "the history of the world"—that brief period of thousands of years measuring the duration of civilization—was only an evanescently short episode in the long course of organic evolution, merely a sliver of the shadow created by the existence of our planetary system. We are grains of perishable protoplasm. Spirit and Matter are united, of this Haeckel was sure. And so he took these lines of Goethe's as the epigraph for his book *The Riddle of the Universe*: "What greater thing in life can man achieve / Than that God-Nature be revealed to him?"

From his mother, Charlotte, he inherited a capricious and feeling heart; from his father, Karl, a curious intellect. Karl was a jurist, son of a jurist, privy council to the Prussian court at Potsdam, where Ernst was born in 1834. His one brother, Karl, followed Papa into the profession, but for him it was the tales of Humboldt that pulled at the heart. The poems of Goethe and Schiller, too, read to the boys by their mother. Theirs was a proper German education.

New winds were blowing through the nation. With Napoleon defeated and the *volk* awakened, liberals like his father held up hope for unification. Down with the petty princes who had carved our land into private fiefdoms. Down with the clerics who peddled revelation rather than rational law. Ernst had a tutor then, with whom he would walk daily through his mother's well-tended garden, learning about the order imposed on nature by Linnaeus. With him he discussed the romantic theories of Schelling, Oken, and Hegel.* As a young boy he was torn between two images of the world: one, the outer world of nature—a vast, ordered museum, established perhaps by God but illuminated by the clear light of science—and the other, an inner world of constant change and metamorphosis, cataclysm and conflict, hidden meanings and buried light. Poets and artists had turned the Enlightenment legacy inside out: The outer world was but a window to an inner landscape. The deeper that research penetrates, the further natural phenomena are reduced to mathematics, the greater grows the temptation to seek the last cause of all things in mechanical blind natural law. And yet how empty, how superficial, how insipid is this conception, and how unsatisfied and comfortless it leaves the soul.

As Haeckel grew older and more independent, helped by the guiding hand of Goethe, he began to see that romantic nature philosophy could accommodate each vision without giving up on the other: substance and energy, abstract theory and meticulous observation, art and science, man and nature, even the old/new philosophy of

* Friedrich Schelling (1775–1854); Lorenz Oken (1779–1851); and Friedrich Hegel (1770–1831), together with the slightly older Johann Gottlieb Fichte (1762–1814), were a group of German philosophers influenced by Immanuel Kant (1724–1804), who developed what the English-speaking world came to call German *Naturphilosophie*, an approach to nature in which a unity of matter and spirit is assumed, and all things have a purpose.

transformation. And he began to abandon God in the heavens, slowly replacing him with God in nature. Goethe wrote and he concurred: "You must when contemplating nature / Attend to this in each and every feature / There's no outside and no within / For she is inside out and outside in / Thus will you grasp with no delay / The holy secret clear as day."

In those days botany was prime in Haeckel's interest. Matthias Jakob Schleiden's* book on the life of plants introduced him to the idea that fossils were the ancestors of the creatures that now climb and crawl the earth, brought about first by mysterious bygone physical forces and later by chemical forces, probably in the tropics. He was beginning to develop the gaze of what had recently been coined "a scientist." Already at a young age, such ideas were planted firmly in his mind.

The failure of the revolution of 1848 dashed his father's hopes, and within a few years he retired. The dream of Germany overseen by an enlightened parliament and ruled by a single king crumbled together with the Frankfurt Parliament.** Shortly afterward, Haeckel's father sent him to Würzburg. It didn't take long before he discovered a complete and utter revulsion; there was no way that he would ever practice medicine. Father stood firm, and so Haeckel matriculated, but his spirit was searching in other realms.

From Goethe he had learned that science and poetry and art all had the very same function: to reveal from under the unending variation an underlying relationship of parts to the whole. At the same time he discovered that he had a true talent: Peering down the microscope with one eye, he could simultaneously draw intricate tissues and organs with the other. So enamored had he become that in a letter to his parents describing his forays into nature, he called his indispensable microscope "my wife." There was a peculiar power of attraction for him in the strange fact that the cell is the origin and

---

* More on this later.

** The first freely elected representative of all the German states. The ill-fated parliament's session lasted from May 18, 1848, to May 30, 1849.

basis of all organic bodies. He considered this to be the greatest miracle of creation, about which he was more astounded and pleased than he could adequately express. When he was in nature, all the troubles he suffered during the day were immediately lifted. The same was true when reading or writing poetry. Like nature, he came to believe, poetry raises a man above the dust.

Having drunk from the chalices of von Kölliker, Müller, and Virchow, Haeckel began his journey into independent research. When he traveled to Italy, leaving his betrothed Anna Sethe back home, he did not find it easy at first to settle on an organism. On the island of Ischia, instead, he met a gnomelike man with a Bedouin nose, a poet and painter named Hermann Allmers. Like Goethe before them, they climbed Vesuvius, losing each other in the dark only to reunite just as all hope was lost. Sailing to Capri, they swam nude at the Blue Grotto, where phosphorescent animalcules painted the waters a brilliant sapphire. To ward off the spirits they sang the songs of Goethe and Heine. They slept late and painted landscapes and ended lazy days of sunshine dancing the tarantella drunk. With Allmers, Haeckel had come to know an artist's life, which tore him away from science. When he was back in Germany, Allmers wrote to him:

We were still strangers to each other
We spoke but a moment at the start
And yet soon each like a brother
Revealed the depths of his heart

Allmers had penetrated the nucleus of Haeckel's soul.

And yet. Though youthful and full of folly, he was sober enough to know that the life of a bohemian was not going to earn him a livelihood. And though Allmers may not have been happy about it, Haeckel's betrothed Anna Sethe was awaiting in Berlin. He wasn't sure, too, that his talents as a landscape artist were more than merely ample. After Allmers left, he therefore settled down in Messina at the tip of Sicily, on September 10, 1859, where Odysseus had steered his ship between Scylla and Charybdis. Every morning he would wake and jump in the sea for his swim, then head for the fish market to inspect the early catch. His eyes could hardly believe all the astonishing splendor of animals: siphonophores and medusae and efflorescent swimming sea snails and slugs. The richness was so transcendent that he called it "the El Dorado of zoology," and like the gold buried in that far-off land, it nearly blinded him. Toward the end of November, he finally settled on a class of organisms he could study: little-known, single-celled creatures named Radiolaria. Secreting silica armors of intricate complexity and aesthetic vibration, these jewels now became his own.

By cosmic design, he thought, his beloved teacher Müller, just before he killed himself, had begun to work on this class of tiny marine protozoa, following their earlier discovery by Ehrenberg.* Haeckel could only surmise that his teacher's spirit and his own were drawn to the same manifestations of nature. Müller had produced a short monograph appearing after his death; Haeckel had taken a copy with him to Italia. Now, using a pipette, he placed a single drop of water on a microscope stage, sometimes searching a thousand drops to find a single specimen. He had to work fast to draw the living creatures under the camera lucida, for they could only be kept alive for a very short while. The strongest magnification and precise angling of the lamp light was necessary for anything at all to be perceived, but he persisted. Every morning he was newly amazed at the inexhaustible richness of these delicates, straining to believe that each creature is but a single cell. Some were like grids or broken nets or stars, others like tiny bowls or helmets or bells, still others like medieval armor, or instruments of torture, king's crowns, Egyptian pyramids. And at the urging of these magnificent creatures, slowly the conflict between mind and heart, precision and passion, began to melt away within him. Like Coleridge's ancient mariner, Haeckel had looked into the sea and suffered a transformation of the spirit.

Upon return to Germany with a suitcase filled with species never before seen, he arranged to work on his unique collection at the Berlin

---

* Christian Gottfried Ehrenberg (1795–1876) had discovered the Radiolaria in 1839.

A page from Haeckel's *Art Forms in Nature*, 1899–1904, depicting radiolarians.

Zoological Museum. It was soon evident that he'd gone well beyond his teacher, and so he wrote it all up for his *Habilitationsschrift*, and translating it into Latin earned him a doctorate. The following year there appeared a large two-volume monograph, *Die Radiolarien*—comprising 570 pages and 35 copperplates. Haeckel had doubled the number of known Radiolaria, becoming the most knowledgeable man of science on their kind. He was making a name for himself, finally. But something else entirely had caught his mind.

When Haeckel was preparing his specimens in the Zoological Museum just after his return from Messina, he came across a German translation of a book by an Englishman named Charles Darwin. He knew full well that his examiners regarded it as completely mad, and did not mention *On the Origin of Species* even once in his dissertation. But when he returned to write *Die Radiolarien*, true gold had been placed in his hands. Darwin's translator, an able paleontologist and morphologist named Georg Bronn, had himself offered a theory of transmutation just two years earlier. But Bronn believed that species could change and progress only through the guidance of a divine hand. Sympathizing with his subject, Bronn nonetheless made it clear that Darwin's interesting theory was at present no more than a hypothesis: There was neither evidence to argue for it, or against.

Peering down the microscope at his radiolarians, Haeckel knew this was mistaken. The tiny creatures' elegance and complexity embodied the unity of art and science, and shouted out that Bronn was wrong. They were among the earliest skeletonized life-forms, an alphabet of possibilities. In their astonishing variety he sought the key to the creative powers of nature, and thanks to the wonders of homology the evidence was now in his hands: He could trace genealogical descent in these single-celled jewels all the way back to an ur-organism, showing through the skeleton, and its relation to the central capsule, that there existed relatedness within families, and many transitional forms. Natural selection had brought the entire doctrine of the relationship of organisms to sense and understanding. This had been the singular genius of Charles Darwin. Building on the microscopes devised by Leeuwenhoek, and the "cell" coined by Robert Hooke, Haeckel could now show it to the world.

To his delight, Anna began calling him her "German Darwin-man." And though he started delivering his gospel to his countrymen, the full force of his theory had yet to seep entirely into his mind. He was still beholden to the old *Naturphilosophie*, to his beloved Goethe and his morphology. Once the ur-parent was discovered, he could derive not only existing forms, but also those that could possibly come into being, or so he thought. Knowing where things came from would allow one to know where things were going. Even if the great Darwin may not entirely agree.

And yet. When he was summoned by the Society of German Natural Scientists and Physicians in the autumn of 1863 to the Eastern town of Stettin, he arrived with a sublime, almost ordained, sense of purpose. He was to defend Darwin's theory of evolution, and there was fire in his eyes when he stepped up to the podium.

By all means, he told his audience, Darwin's proposal was not novel: Lamarck, Geoffroy, and even Germany's own Oken, had known that creatures big and small undergo continual change. Goethe, to be sure, had expounded at length on how later forms were genealogically related to those that came before them. But it was Darwin, thanks to his idea of selection, who had seen most clearly the perfecting hand of nature. In the struggle for existence, those members of the species will survive if they have greater quickness to escape predators, more strength to withstand attack, better organization of structure, precise sense organs, larger stores of energy, or a more abundant libido. The three-fold parallel between embryological, systematic, and paleontological development of organisms was all the proof one needed; ever climbing lineages made it plain that the history of organisms manifests the law of progress. Haeckel's closest confidant at Jena, the linguist August Schleicher, had dedicated a book to him in which he'd fit all languages on a tree. Now Haeckel could see that what is true for languages is true for life generally. Higher and higher the tree's branches would grow, elevating the human races. Aptly, he named it "a progressive metamorphosis."

That winter, Haeckel rubbed his eyes at the arrival of a letter postmarked from London. Having received Haeckel's *Radiolarien* volumes, the letter-writer called them "the most magnificent works which I have ever seen." Emboldened, Haeckel sent him newspaper clippings from the event the previous fall in Stettin. "You are one of the few who clearly understands Natural Selection," came the reply, signed Charles Darwin.

Haeckel wanted to cry with joy at his hero's praise, so his letters tell us. But he was numb, devoid of feeling. Anna had died a few short weeks before. There was no point left to anything. Devastated and nearly deranged, Haeckel's universe had turned as dark as a plum left to rot on a winter tree.

6.

CELLS AND EMBRYOS

Antoni van Leeuwenhoek was beside himself. The "gentlemen" over at the Royal Society of London seemed not to take his observations seriously. They wanted more explanations, more proof. But he knew what his eyes saw—those millions of "animalcules" in his sperm—moving forward "with a snakelike motion of the tail, as eels do when swimming in water." So what if he had to employ artists to make his illustrations, since he could not draw for toffee? So what if he used common terms in Dutch, rather than scholarly Latin, to describe what he had seen? Vainglorious and conceited, the cloth merchant from Delft saw no reason why he should let observers examine his microscopic instruments. Who were they to doubt his genius?

Still, the Royal Society did publish his letter on animalcules seen through a single lens microscope in a drop of water. And in that same year, 1665, the English polymath Robert Hooke had seen cells too—and named them—though dead ones in a sliver of cork slipped under his compound two-lensed microscope, not anything as bizarre as those reported by Leeuwenhoek. Even fifteen years later, Leeuwenhoek was still nursing the wound when he finally got around to reading Hooke's masterpiece *Micrographia**: "It has often reached my ear that I only tell fictitious stories about the little animals," he wrote the Englishman. And yet, by the end of his life, in 1723, Leeuwenhoek could prophesy with confidence: "Nay, we may yet carry it further, and discover in the smallest particle of this little world a new inexhausted fund of matter, capable of being spun out into another universe." The cell was now a fact. Who knew what further mysteries lay within.

In the eighteenth century, many botanists and zoologists would turn their eyes at cells under their microscopes. But it was a Frenchman from Vaucluse who, more than anyone, began to unravel what cells were all about. François-Vincent Raspail had a mind that was fiercely independent. A self-taught botanist and chemist, as well as an attorney and physician, he joined an illegal, left-wing secret society during the Revolution of the 1830s. Thrown in jail by the authorities, he trained fellow inmates in matters of sanitation, hygiene, and antisepsis before becoming a candidate for the presidency of the French Second Republic in 1848, coming in fourth. Although Raspail was tried and convicted for his involvement in an earlier coup attempt, the winner of the election, Louis-Napoléon Bonaparte, commuted his sentence and exiled him to Belgium. Eventually returning to Paris, he became an elected deputy in Parliament. All the while, he published learned scientific papers on an impressive array of subjects, most important of which were his studies of cells.

Raspail presaged much of modern cellular biology by combining seer-like intuition with careful observation. Cells are autonomous, he claimed, constructing themselves from building blocks in their

---

* The full title of Hooke's 1665 book was *Micrographia: Or Some Physiological Descriptions of Minute Bodies Made with Magnifying Glasses with Observations and Inquiries Thereupon*. It contained thirty-eight minutely detailed drawings of everything from ants to mold to fleas. The English diarist Samuel Pepys called it "the most ingenious book I read in all my life."

environment. With walls (later called membranes) that could selectively allow in water, carbon, and all kinds of other elements, they were tiny laboratories in which biochemical reactions helped sustain life. Raspail was a pioneer in theory but also in the use of a technique of freezing cells under the microscope and treating them with dyes. Already, in an early unpublished manuscript from 1825, he wrote an epigram that is astonishing in its perspicacity. In our current-day language, cells are metabolic units responsible for an organism's physiology. But where did they come from? *Omnis cellula e cellula*, Raspail wrote there, without providing evidence, "All cells come from cells."

Today, there is a boulevard and a metro station named after Raspail in Paris, but when he made that leap of imagination, cell theory was just taking its first steps. Stacked against it was the medieval worldview of preformation. By imagining tiny humans folded into drops of sperm, going all the way back to Adam and Eve, preformationists had managed to avoid giving any real account of embryological differentiation. They also held that organic beings weren't reducible to chemistry and physics. The observing eye and pious mind, they claimed, made that clear.

But how could a creature as complex as man or woman arise from a fertilized egg that seemed homogenous? If life came preformed, why couldn't we see it? A clue finally came thanks to another student of Haeckel's old teacher, Johannes Müller. Robert Remak was born in Posen, Prussia, to a family of orthodox Jews. Remak's research had already made him a leader in the study of cells. Among other things, he had discovered unmyelinated nerve fibers, and the nerve cells in the heart that now carry his name. Despite these accomplishments, he'd been refused a full professorship at the University of Berlin because he was Jewish. One day Remak saw something almost incredible, but something he had been waiting to see. Placing a red blood cell from a chicken embryo under the microscope, his eye opened wide: The cell suddenly quivered, grew bigger, and split in two. Remak knew that a German botanist named Hugo von Mohl had seen the same doubling at the tips of the roots of plants. Observing frog eggs to prove the phenomenon universal, he now confirmed

Raspail's unpublished conjecture. *Omnis cellula e cellula* was a strike against spontaneous generation greater than Francisco Redi's, since Remak had actually shown under the microscope that life comes from life, rather than wind or mud or rot. It was also a great blow to the doctrine of preformation. Reluctantly, the University of Berlin made Remak an assistant professor. He was the first Jew to hold such a position, a mark of growing integration, of progress. His grandson, a brilliant mathematician, would perish in Auschwitz in 1942.

Though Remak would receive little credit for his discovery, beyond his long-overdue professorship, two other students of Müller's would become celebrated for their work in the new, microscopic process of cell division. One of them, Matthias Schleiden, had started off as a lawyer but was so miserable in his profession that he attempted, unsuccessfully, to put a bullet through his head. Having survived, Schleiden abandoned the law for his true vocation, botany. By this time, major improvements had been made in tissue-preservation techniques and microscopes were considerably more powerful than those used by Leeuwenhoek and Hooke. On a cold October evening in Berlin in 1837, Schleiden met for dinner with his friend Theodor Schwann, who was then assisting Müller in his work on nerves, muscles, and blood vessels.

Schleiden told Schwann that whatever part of the plant he looked at, he had seen the same thing. Whether stem or leaf or petal or root, the tissues were aggregates of individual cells, entirely distinct and autonomous. Schleiden knew the work of a Frenchman named Marie François-Xavier Bichat who showed that there were at least twenty-one different kinds of tissue in a human body, giving rise to the new science people called *histology*. But while the cells might look different from one tissue to the next, there was a unity beneath the diversity that he could now clearly sense. Cells were units making whole, independent lives within a body. And all of them were variations on a theme.

Schwann looked at his friend in amazement. The same was true, he said, of the animal cells he'd examined in amplification. A blood cell or a liver cell or a cell that came from the heart were all essentially the same. Each had a perimeter, and each had a nucleus. All were leading double lives, manning their own borders while building whole organisms. Although Swammerdam and Hooke and Leeuwenhoek and Bichat and Raspail had already observed the same bricolage-like aggregations before them, Schleiden and Schwann now took a dramatic, extra step.* Cells weren't merely passive structures, they wrote in separate treatises. Cells were the living, self-governing building blocks that made up all creatures. Laying down the first two postulates of what became known as "cell theory," they announced: (1) All living organisms are made of one or more cells, and (2) the cell is the basic structural unit organizing all forms of life.

But if cells were so fundamental, where did they come from? How had the first fertilized egg become two and then many, bringing about the multifariousness of life?

Before he died, or killed himself, their mentor, Johannes Müller, thought he had the answer. Chemicals often created structures by a process called crystallization; perhaps the vital fluid giving life to a cell did the same. Schwann took a closer look at the analogy and found it sorely wanting; crystallization of a vital fluid led to all kinds of paradoxes that made absolutely no sense. Müller was transitional, a man caught between Romantic and modern times. Perhaps his somewhat mysterious death at age fifty-seven had something to do with his inability to reconcile purposefulness with inert protoplasm, a planned world with a world beholden to the dictates of matter.

This was where cell theory stood when Haeckel's stern professor from medical school came onto the scene. Rudolf Virchow was short, bespectacled, and formidable. When rebellions broke out across Germany in March 1848, he wrote articles in support of the Revolution, treated the wounded, and mounted the barricades, pistol in hand. A few months earlier he had traveled to Silesia to figure out how the famine there had led to a terrible epidemic of typhus. Soon it was clear to him: It was not just the disease agent that was responsible. The

* For almost every novel observation or theory in science, there are antecedents. In this instance, the botanists Pierre Jean François Turpin (1775–1840) and Franz Meyen (1804–1840) stand out.

nation was bleeding from years of misrule, autocracy, and neglect. The people were treated like dirt, spat upon by those in power. "The body is a cell state in which every cell is a citizen," he wrote, grasping that pathology and politics were linked. When the rebellion was quashed, Virchow was branded a liberal. Kicked out of his post as a doctor at Berlin's Charité Hospital, he was sent to Würzburg and ordered to stay out of trouble.

It was there that Haeckel first encountered Virchow, having himself been sent to medical school by his father. By this time, Virchow was already recognized as one of the greatest scientists of his age. As a young physician, treating a female patient who complained of exhaustion and exhibited an enlarged spleen and bloated abdomen, he had noticed an inordinate number of white cells in her blood, and named her disease *leukemia*. Setting out to discover where all the white cells had come from, Virchow grew positive that it had nothing to do with crystallization. Instead, what Hugo von Mohl had seen in plants, and Robert Remak in chickens and frogs, likely also occurred in humans. Eventually, Würzburg proved too small for Virchow's talent. Summoned back to Berlin, he published a book that appeared a year before the *Origin of Species*. Virchow was about to change the world, just like Charles Darwin.

For centuries doctors had claimed that disease was due to an imbalance between the body's four humors. Or that it resulted from bad air that carried contagion, or from the breakdown of entire tissues or organs. A year earlier in Paris, the Frenchman Louis Pasteur had initiated studies that would lead to the treatment of beer and milk, a process that would be named pasteurization. Microorganisms infecting animals and humans, Pasteur assumed, were the agents responsible for disease. But what Virchow argued in *Cellular Pathology* in 1858 took things one step further. However they are contracted, diseases always first begin within a cell.

Virchow had figured out that life was, in essence, cellular activity. Examining the tumors and blood of many patients, he concluded that what the Greek physician Hippocrates had called *cancer* in the fourth century BC on the Island of Kos was the result of cell division gone mad. Microorganisms, on the other hand, were invaders interfering with cells that made a healthy body. Now he could generalize and add three more postulates to Schwann and Schleiden's cell theory. Not only were all living organisms made of one or more cells (1), and not only were cells the fundamental units of structure organizing all forms of life (2). Now it was clear to him that all cells come from cells (3), that health is the function of cellular physiology (4), and that disease is the result of healthy cellular physiology gone wrong (5). These five postulates became the foundation of modern biomedicine.

It was all very momentous. Science had discovered where cells came from, and that they were living, breathing agents central to every aspect of creatures' lives. Now cell theory could be applied to the mystery of *generation*. Already, the catchall word had been disappearing from scientists' tongues, replaced by the more specific terms *reproduction*, *conception*, and *development*. Thankfully, Lamarck had coined a name for the new science of life, in 1802—*Biologie*. But there were basic questions that remained unanswered. Chief among them were: Can life come from nonlife? How does inheritance work? And do creatures come ready-formed or develop gradually?

The solution to this last conundrum was intimately tied to the old argument about sex. In an experiment that seemed designed to humor children, the eighteenth-century Italian priest Lazzaro Spallanzani dressed male frogs in tiny little diapers and showed that when their sperm was blocked, the female's eggs would never fertilize. When he dipped a paintbrush in the sperm, on the other hand, and applied it to the eggs, tadpoles materialized. It was the world's first demonstration of *in vitro* fertilization.*

But neither Spallanzani, nor anyone else, had ever explained how precisely life develops. That was left to a young aristocrat from the Governorate of Estonia named Karl Ernst von Baer. In the winter of 1827, for the first time in history, his eyes fell upon a mammalian egg, of a dog. Philosophers and microscopists were finding it difficult to

* Spallanzani would later perform the world's first artificial insemination, in a dog.

explain why so many souls had to be lost in the act of procreation, or why "monsters" were occasionally born if divinely ordained life was preformed in the male seed. Now von Baer came down with what seemed to him the only possible explanation: The egg had defeated those who'd thrown their lot in with sperm.

Von Baer argued that all sperm did was to awaken the ovum. To him, Leeuwenhoek's "animalcules" were just parasites, which he christened *spermatozoa*—animals of the semen. Still, he took a closer look at what happens after their union. And what von Baer found went against the doctrine of preformation. To preformationists, familiar images like the peeling back of a bud in early spring or the opening up of a bean to find folded leaves and flowers within in early germination—even digests or compendia that carried in miniature all the content of an original book—all argued that development was a mere unfolding of what was already present. At the same time, the discovery—by the Genevan naturalist Abraham Trembley in 1744—of the curious ability of a freshwater creature called a hydra to regenerate after being split into two raised a problem: Had the hydra's soul, not just its body, also split in two? Preformationists had hit a wall.

Building on the work of those who had considered the problem before him,* von Baer attempted to explain development in material, rather than nonmaterial, terms. What he showed was dramatic. Mammalian embryos were assembled through a process in which the egg gradually turned into a ball with three layers: an *ectoderm* giving rise to the nervous system, skin, nails, teeth and hair; an *endoderm* generating the lining of the gut and internal organs; and a *mesoderm* that brought forth muscles, bones, and hearts. None of these structures had existed previously. Rather, like the babies they'd eventually give rise to, they came into being gradually.

---

* In particular, in Germany, Caspar Friedrich Wolff (1733–1794), who observed the stage-by-stage development of different organs like the heart and intestines, and Johann Friedrich Blumenbach (1752–1840), who coined the term *Bildungstrieb* (formative drive) to describe the epigenetic development of plants, animals, and humans.

But if individual life came about gradually, was life writ large also changing? Were entire lineages and orders and families in any way progressing over time? Already in the 1820s, a German and a Frenchman defined a law aligning embryology with comparative zoology. The Meckel-Serres Law of Parallelism, as it became known, stated that "a higher animal in its embryological development recapitulates the adult structures of animals below it in the linear 'scale of being.'" What this meant was that new life-forms somehow reenacted all the successive life-forms that came before them. Von Baer took a closer look but wasn't duped.

Embryos weren't going through the adult structures of those "below" them. Rather, they were advancing from the general to the specific. Starting out with characteristics shared by all members of their archetype, the growing embryos fine-tuned themselves, gradually assuming the uniqueness of their species. A vertebrate was a vertebrate from the very beginning, and never went through the stages of a mollusk, say, or a centipede. As an embryo it always started off as a general form of a vertebrate, and proceeded from there. If it was a human embryo, for example, it would assume the shape of a mammal, then a primate, and finally a human being. Exactly how this happened had to be explained by the mechanics of development, a science he gave the unwieldy German name *Entwiklungsgeschichte* (the history of development). Later it would be dubbed the equally tongue-twisting *Entwiklungsmechanik* (developmental mechanics).

But von Baer turned a cold shoulder to evolution. Taken to its logical conclusion, he thought, it would mean that a developing fetus could pass from one archetypal form into another—a proposition he found absurd. Clearly, the evidence didn't support what Meckel and Serres had called "recapitulation," namely the fetus moving through adult forms of lower organisms in the same archetype. Von Baer was conflicted. He stood at the helm of cutting-edge knowledge, but Aristotle was right: Development is a hard nut to crack. Though he had given it its name, von Baer wasn't even sure that his own science could entirely describe development. Following Kant and Goethe, he held that creatures possessed some sort of internal drive, some inner purpose, without which all the mechanics in the world were useless. If you had to give it a name worthy of its ancient and early-modern roots, his philosophy was a form of teleological epigenesis.

But if egg and sperm could produce such a miracle, what exactly were they? Thanks to Schleiden and Schwann, it now became clear that egg and sperm were cells, variations on liver and heart and brain cells, whose own specialty was producing the next generation. If that were true, despite the differences in size and motility, maybe egg and sperm were equivalents. And maybe the reason they came together was simply to make a new cell. Embryos weren't preformed; they were created by dividing cells. Cells dividing and dividing and somehow turning into all the different kinds of cells a creature needs. How this occurred would entail figuring out the relationship between development and heredity, the mountain that everyone since the ancients sought to climb.

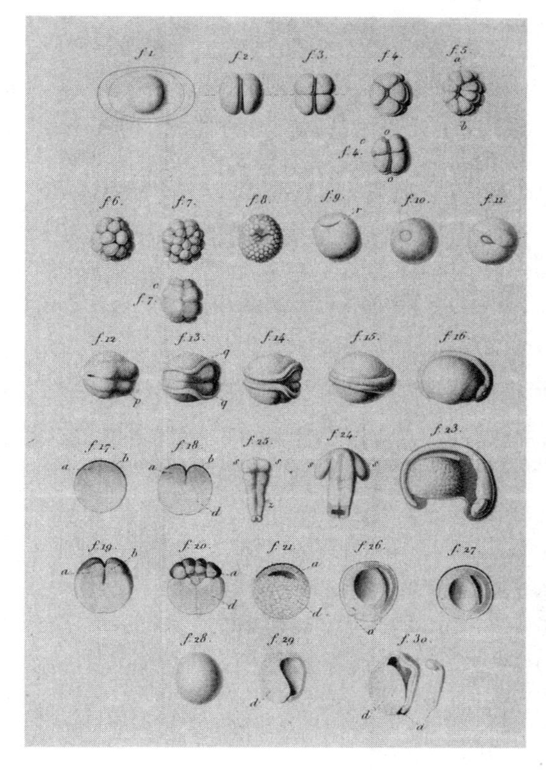

A nineteenth-century (1835) drawing of frog and salamander embryos, showing the initial divisions of the eggs, and the creation of layers. The author, the Italian Mauro Rusconi, was actually a rival of von Baer's. Despite improving microscopes, neither could agree how precisely the layers were born.

Today, we know that von Baer's three layers—the endoderm, mesoderm, and ectoderm—are indeed the repository of cell layers that form the body. Dividing and differentiating, each cell type in each layer will create specific tissues like nerve and muscle; particular structures like toes and elbows; and entire organs, like hearts and livers and brains. As we shall see (Part 3), this involves a symphony of regulation. A regulation that happens at the level of genes.

In the mid-nineteenth century, genetics had yet to be discovered. Instead, the great achievement of the day had been to show that life was made of cells behaving in different ways. Health and development could be explained by cells growing and dividing. Death could be explained by cells getting diseased, destroyed, or just old. And reproduction and conception could be explained by cells coming together and uniting. It was a magnificent achievement, even as the workings of heredity remained unknown. Scientists like von Baer had begun to show how creatures are made gradually. Step-by-step, hour-by-hour, cells turn us into who and what we are.

And step-by-step, hour-by-hour, Yaeli's pregnancy is advancing. When the kids first heard they were going to have a baby sister, they were thrilled, but also a little confused. With their mommy's belly flat for all those months, it sounded and looked to them like a made-up fantasy. But now, as the bump graduated into a hill, and then surprisingly into a mountain, things have started getting real. We're halfway there, and all four of us know that more or less everything is already in place. The three layers were laid down long ago, and sprouted legs and arms and eyes and kidneys. Connecting tissues joined nerves to muscles, allowing a little brain to execute commands. Even the appendix has by now sprung into action, making hormones that keep our baby's internal environment in homeostasis. Yaeli says she's starting to feel her inside, like gentle waves of water. The doctor says our daughter is about the size of a small banana.

One evening, Yaeli is on her back in bed looking like a beached whale. Snuggling beside her, we decide to sing to the baby. Curly-haired Abie jumps up first and belts out "Baaaaby shark tootoo tootootootoo, baby shark tootoo tootootootoo . . . ," seven times, before falling down on a pillow, spent. Next, Shaizee sings a beautiful old Hebrew song about a Hyacinth, *Shir Layakinton*, pressing her mouth to her mommy's belly. Finally, I deliver a rendition of Gershwin's "Summertime" so out of tune that everyone plugs their ears. When Yaeli cracks up first, we all laugh and tumble, assured that change will bring us happiness.

But none of us has a clue what lies ahead.

7.

DEUS SIVE NATURA

Haeckel returned from Nice burning with the discovery of "the most charming and beautiful of medusae," the one he'd named for his beloved with her curly locks of blond hair. Still crushed by Anna's death, his spirit decimated, despair would nevertheless soon take on an unusual form of passion. To Darwin he wrote that he was hardened against the blame as well as praise of men, impervious to external influences, that he now saw but one and only one goal in life: to work for his theory of descent. He would support it. He would perfect it. He would help bring it to the world. This would be his memorial to Anna.

It took time to finally sit down to write his magnum opus. At first he sought distraction, organizing a trip with his students to Helgoland, where he had serendipitously met his teacher Müller a dozen years before. Then, unable to resist, he visited his old friend Allmers. But back in Jena in the fall of 1865, he began to lecture on Darwinism to auditoriums brimming with aspirants, and returned to his readings in Kant. Slowly, his sorrows were forged into a spear as sharp as that of Ares. Within the space of one year, he produced two volumes, more than a thousand pages all told. He called it the *General*

Morphology of Organisms. Hopelessness had morphed into zeal. Nature-Anna was now his muse.

A paean to the three brightest stars in the firmament of the great men of the century—Lamarck, Goethe, and Darwin—Haeckel set about in his book to describe the process by which organisms take form. To do so, he would need to go beyond his hero Goethe's abstractness—to propose and test different theses that would render descent theory bound by rules. He littered his book with nearly 150 laws of form and formation, following Darwin in his chapter on "laws of variation." It was high time someone made a science of the disparate observations that made up the mess people called morphology, or so he thought.

The reception was not easy. Experts in developmental mechanics attacked him for not sufficiently following scientific principles of cause and effect, succumbing to a speculative past rather than adhering to the here and now. Such men performed careful experiments on fertilized eggs and embryos, but to Haeckel's mind had forgotten about evolution, alas. His own students, Roux and Driesch, were of this bent, but his greatest adversary was His the Swiss, who complained that he had rudely interjected descent theory into the realm of professional embryology, where it did not belong.* Haeckel had no difficulty rebutting him, since by then he'd strongly rooted his philosophy: Ontogeny may be a process that takes place within an individual, he was sure, but ontogeny has a history that goes back in time long before the individual is conceived. The union of an egg and a sperm and its aftermath should be described not by sterile physics but by a historical natural science. To stand a chance to fully comprehend how life develops, it was paramount to recognize that matter had gone through a long Darwinian process of adaptation and inheritance.

More than this, force and matter were not different, rather they were two manifestations of the very same oneness. The living and

* Wilhelm Roux (1850–1924), Hans Driesch (1867–1941), and Wilhelm His (1863–1934) were all biologists concerned with development, whose work will be further discussed in Part 3.

nonliving were indistinguishable, periodically morphing into each other. Kant had ordained that the task of a biologist was to reduce as much as possible the teleological properties of creatures to mechanical descriptions, a task he believed could never be entirely achieved. But despite his greatness and what he and his idealist acolytes preached, Haeckel came to believe that their imagination had reached an impasse. Religious critics would moan that he had reduced life entirely to a materialist metaphysics, reviling him for disenchanting their holy world. More irritating for Haeckel were men of science who accused him of suffusing his theories with *telos*. How they misunderstood him! Mind and matter were one, indivisible, indissoluble. With Darwin's descent theory to wield as his weapon, he could vitalize matter as much as he could materialize life. Haeckel was no longer confused: *Naturphilosophie* that devolved into fantasy was crass, but neither was lifeless physics tenable. Forging a middle way, he would prove the skeptics wrong who believed there could never be a "Newton of a blade of grass."

This was a battle, and called for a military state of mind. He determined to decimate those imbeciles grubbing after mundane facts, those faux philosophers stumbling through a conceptually impoverished dream life, as if slumbering in a half-sleep. Not even his allies liked this. Though Darwin approved of his "wonderfully clear and good" exposition of the fact and cause of divergence of character, as he wrote to him after receiving the two *Radiolarien* volumes, in treating those who had yet to accept descent theory, Haeckel had perhaps been too severe. Darwin expressed worry that his German friend's strong language would lead readers to take the side of those whom he attacked. Haeckel replied politely that even his best friend* had admonished him in the same manner, providing the excuse that he had written such words while suffering from an extraordinary bitter attitude and nervous state prompted by the death of Anna. There was

* Karl Gegenbauer (1826–1903) was a leading comparative anatomist and evolutionist at Jena and an important teacher and intellectual influence on Haeckel.

truth to this, but Haeckel's feelings were entirely genuine. Enough with gentle exposition, he thought, and pandering to inferior minds. He would battle the enemies of Darwin like a mother lion protecting its cub.

Evolution was a fact, of this there could be no argument. Nor could it be doubted that natural selection was its primary cause. How the variations that feed natural selection arise Haeckel could not say with complete confidence. But he ventured guesses, just like Darwin.* Neither man knew how heredity worked, only that it could operate indirectly through a silent change to the parent sex cells, or directly by the inheritance of acquired traits. It was true that species formation was dependent on the mutual relationships of organisms to one another, a science Haeckel had duly dubbed ecology. It was true, again, that geography, topography, and behavior mattered, a science we call biogeography and that Haeckel then called *Chorologie*. But above all, what was true was that natural selection would bring to bear a slow but constant improvement. Life didn't have an intrinsic tendency toward improvement, as blind believers in *telos* professed. But, struggling in a competitive environment, life was bound over time to progress.

With this in mind, Haeckel showed that metazoan animals arose from a hypothetical single-celled progenitor, an ancient creature he called *Gastraea* that had aggregated and then adapted by making a division of labor among its parts. What a biological individual is was therefore a crucial question, and Haeckel made a tripartite division in the hope of clarifying the problem: There were morphological

* While Darwin spoke of seedlike "gemmules" in different parts of the body that could be changed by use and disuse and then transmit that change by flowing through the blood to the sex organs, Haeckel spoke of "plastids," particles in the protoplasm of cells that would vibrate in different frequencies based on how the body was being used and then transmit those vibrations to the sex organs—an idea that was ridiculed by many contemporaries, including Virchow. With little evidence to support them, both men also believed that direct action of the environment on the sex cells could alter them, producing variation upon which natural selection could work.

individuals, defined by their indivisibility, rendering hydra a challenge to biologists. There were physiological individuals, defined by their talent for self-maintenance. And then there were genealogical individuals of different sorts—the reproductive cycle of a person, or bion, from conception to maturity; the species, or collection of bions over an extended time; and finally the "phylum," a name he invented to denote all the genealogically related species who sprang from an original parent stem deep in the evolutionary past. As much as Haeckel sought one, no conception of an "absolute individual" was therefore possible. An "absolute individual" could only remain a fantasy.

Still, Haeckel soon saw that through graphic representation he could unite these different senses of individual to show the power of evolution. Darwin's translator into German, Bronn, had drawn twigs as abstract examples of how species might be represented evolving in relation to one another, without placing any actual life-forms on the various branches. Darwin did the same, in the sole illustration in his *Origin*. But Haeckel, for the very first time, drew fully realized stem-trees to depict the evolution of real Kingdoms. He got the idea from his friend August Schleicher in Jena, who had applied it to languages. In the fifth edition of Haeckel's great book, Darwin praised the talent he'd brought to bear on what he called "phylogeny." Haeckel's trees had made evolution come alive.

When Haeckel looked at them, and then again at the embryos he dissected, a pattern came to view that he couldn't ignore. In their development, physiological individuals displayed forms characteristic of the chain of their ancient ancestors, a chain that through the depths of time reached back to the origin of life. Others had spoken of this, like Étienne Serres and Johann Meckel, even Germany's great Lorenz Oken. Of mammals Oken had written that "the fetus, through the course of the several forms of its existence, is the whole animal," by which he meant that it passes through stages comparable to the polyp, plant, insect, snail, fish, and amphibian, before finally reaching its mammalian form. Karl Ernst von Baer, the Estonian, had nevertheless rejected this notion, and among German biologists

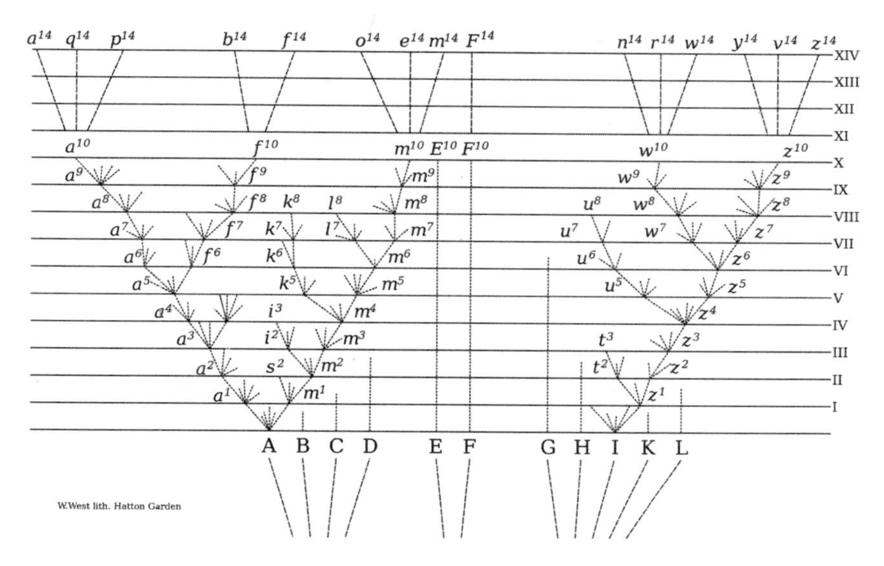

Above, Darwin's simple and abstract evolutionary tree, the only illustration in *On the Origin of Species* (1859), compared, below, to Haeckel's detailed stem-tree of vertebrates from *Generelle Morphologie* (1866).

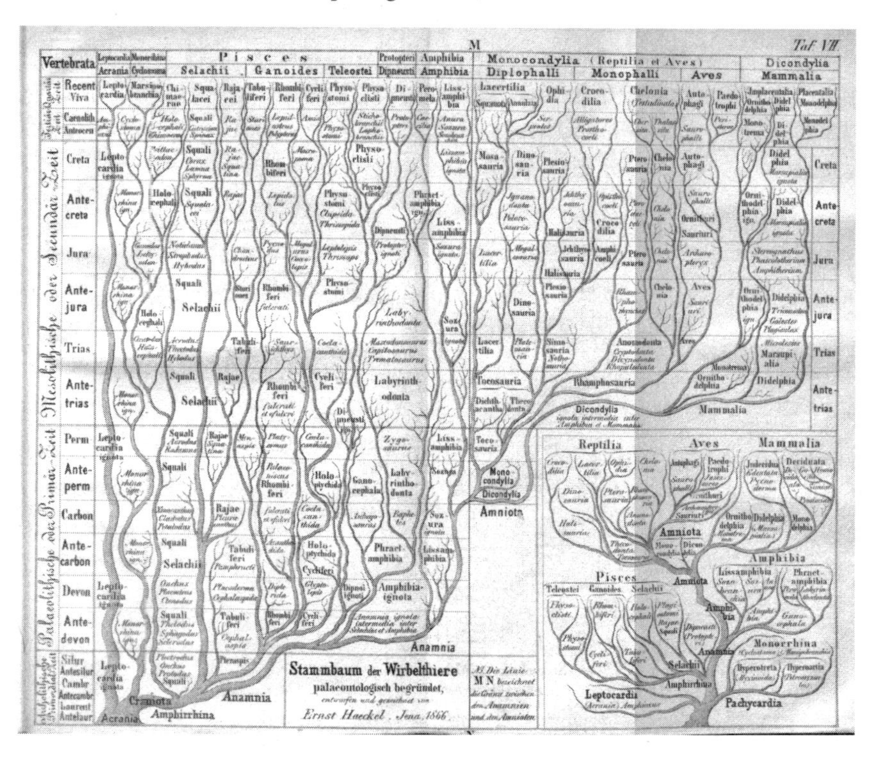

von Baer's word carried weight; across the Channel in England, even Darwin's nemesis Richard Owen took the bait.

But Haeckel now saw clearly how the individual repeats during the short course of its development the most important of those changes in form that its ancestors had gone through during the long course of their evolution. Darwin saw, likewise, how new adaptations sequentially acquired by a species over time would be preserved in their embryos. The Estonian was correct that a vertebrate embryo develops from the more general to the more specific, but both Darwin and Haeckel grasped what he failed to see: that the general vertebrate elevated by him to the status of "archetype" was merely a once-existing adult progenitor of the vertebrate line. Observing a human embryo develop in time, the eye could see first a single cell, then a larger mass of structured cells, then a body with a tail and gills, then webbed fingers, snoutlike nose, and finally round head and upright posture that are the mark of our species. Nothing made more clear how we humans had once been amoebas, sea squirts, fish, amphibians, then four-legged mammals. Change came about through modifications being added successively at the end of embryological development. Embryos, like a folded accordion, were therefore a record of evolutionary history. Later Haeckel would call this the biogenetic law, but he already grasped its essence: Ontogeny is the recapitulation of phylogeny.

That summer of 1866 saw the end of the Austro-Prussian War. In August, at the Peace of Prague, Austria capitulated to Bismarck, and at long last Haeckel could travel with greater ease. En route to marine researches in the Canary Islands, he decided to stop in England. With a trembling spirit, on Sunday, October 21, he boarded the train from London to visit Darwin in the village of Downe. When Haeckel entered his home, all he could do was repeatedly shake his hand, having all at once forgotten his English. Darwin's eyes were so kind, his voice so quiet and soft, that Haeckel believed he had before him the kind of noble wisdom of the Greek ancients, Socrates or Aristotle. Darwin offered financial support for an English translation of Haeckel's book, trimmed of invective. Haeckel knew that he had a friend in him, and returned to London with his heart overflowing with love.

On Tenerife a fortnight later, he met his students and a *Privatdozent* from Bonn.* Like Humboldt over fifty years before them, they embarked to scale the 12,000-foot volcanic mountain, Pico del Teide. While the others faded, alone with his guide Haeckel made for the crest, but blood came running like a river down his nose, and he fainted. After the guide revived him, he fainted as well, and Haeckel now revived him. No matter what, he would not descend defeated, and on shaky legs they scaled the final three hundred feet. Blissfully, with the wind in their hair at noon, they viewed the vistas of the bay from the peak at last.

It was on the nearby island of Lanzarote that the men finally went to work, the *Privatdozent* on worms and echinoderms, Fol on ctenophores and mollusks, and Miklucho on sponges and fish. Haeckel's puerile students, he felt, were harmless enough, but inexact and inefficient. For himself he saved the more interesting radiolarians and siphonophores, a delicate order of colonial hydrozoa as intricate and beautiful as bouquets of flowers. The siphonophores' anatomy had already been drawn by others, but using Darwin's theory of evolution and the biogenetic law, Haeckel now showed how their lineage could be worked out with power and grace. His monograph on the subject would later win him a prize, and the experiments included would foreshadow what we now call "evo-devo."** But by March, Haeckel had his fill. Traveling through the Straits of Gibraltar to Algeciras in Cádiz, exhausted, he trained through Paris to Leipzig and caught the coach home.

Now it became clear: Haeckel was lonely and would need a companion. As he later wrote to Darwin, he therefore married the daughter of his best friend's academic successor, a sensible and

* The students were Hermann Fol (1845–1892) and Nikolai Miklucho-Maclay (1846–1888), and the *Privatdozent* (a former student) was Richard Greeff (1829–1892).

** To be discussed further in Part 3.

Haeckel, standing, with his student Nikolai Miklucho on Lanzarote (1866). *U. Hoßfeld et al. (Hrsg.): Vorlesungen über Zoologie von Ernst Haeckel.*

cheerful girl who cared little for science named Agnes Huschke. As much as he willed his second marriage to be successful, he knew in his heart that his Anna was irreplaceable. Just days after the hasty union, he had thoughts of taking his own life. As Agnes came near to completing her ninth month of pregnancy with their first child, Haeckel set off with Allmers for an excursion in the Italian Alps. To her dismay, as she wrote to him in deep sorrow, his poor wife had noticed that he took with him a photograph of Anna. She called him a hard-hearted professor, and, of course, truth was on her side. But be that as it may, Haeckel's heart belonged to Anna. As it happened, he only made it home in time for his son's birth because he'd contracted a tooth infection in Bolzano, and needed to cut his Italian jaunt short.

Besides his true emotions, what became clear upon return to Jena was that Haeckel's great book on general morphology was about to sink into oblivion. His prose was arcane and his subject very technical, and no more than a few dozen quibbling scientists had read his offering, and they, too, according to him, not particularly well. He would need to popularize his book if he wanted to bring his philosophy to a wider public. And so he published an illustrated book, *The Natural History of Creation*. It would go through twelve editions by the time Haeckel died, teaching more people about Darwinian principles than even Darwin himself. There was no planner outside, he argued there, only a spirit within. The universe lacked purpose, but life was rising.

In the Second Reich, Haeckel's message became a sensation, a quiver of progress drawn by the rising German nation and pointed at an antiquated Church. When men called his philosophy a veiled form of atheism, he let Goethe's Faust reply for him, as any good German would:

> *Fill your heart [with the wonders of nature], so great are they,*
> *And when you are completely blessed in the feeling,*
> *Call it what you will,*
> *Call it happiness, call it heart or love, call it God!*
> *I have no name for it.*
> *The feeling is all;*
> *The name is noise and smoke*
> *That clouds over the heavenly radiance.*

If God was anywhere or anything, he was the causal law that could be comprehended. The philosopher from Amsterdam, Spinoza, had described this in just three words—*Deus sive Natura.** Finally Haeckel was rid of the dualism preached by the orthodox Abrahamic religions, those backward believers, in his eyes, who had torn asunder the material world from immaterial God, the great creator and sustainer. He could fully embrace the sole substance, at once matter and energy, body and spirit, God and Nature. That this substance,

* Baruch Spinoza was born in 1632 and excommunicated by the Jews of Amsterdam, dying in 1677. *Deus sive Natura* translates to "God or Nature."

according to the inviolable laws of physics, could never be destroyed provided solace: Anna would not be dead forever. Perhaps she had already been reborn, as a siphonophore medusa, or a tiny radiolarian. This was the purest form of monotheism and so he dubbed it appropriately.

Haeckel was destined to be a believer. Monism would be his religion, and he its high priest.

8.

AXOLOTL

It was a tragic affair, I knew. James Matthew Barrie, the Scotsman, had lost his older brother David to an ice-skating accident on the day of David's fourteenth birthday. James was only six at the time and would often dress in David's clothes and whistle, imitating his brother to make their mother smile. "Is that you?" he remembered her saying once, as he walked by her room, and having to answer: "No, it's no' him, it's just me."

When he grew older, James moved to London and became a playwright. He met a group of boys from the Llewelyn Davies family one day strolling with their nanny in Kensington Gardens. One of them was called Peter. James wiggled his ears at them, wrinkled his eyebrows, and soon they all became friends. Eventually *Peter Pan* opened on December 27, 1904, at the Duke of York's Theatre in London. George Bernard Shaw wrote that the play by J. M. Barrie, as he became known, was really for grown-ups, that Victorian Britain had suffered through industrialization, and that the island of Neverland appealed to countless Englishmen and -women who lived in cold, alienating cities. Other contemporary critics read the play as an antidote to the century of James Clerk Maxwell and Charles Darwin,

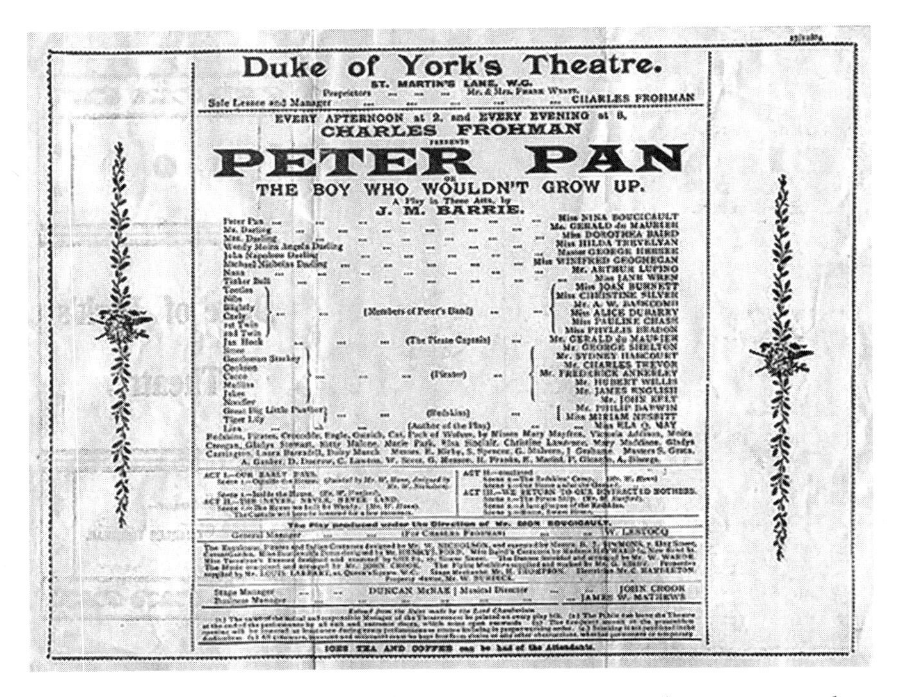

Duke of York's Theatre.
ST. MARTIN'S LANE, W.C.
Proprietors — — — Mr. & Mrs. Frank Wyatt.
Sole Lessee and Manager ... ... CHARLES FROHMAN
EVERY AFTERNOON at 2, and EVERY EVENING at 8,
CHARLES FROHMAN
PRESENTS
PETER PAN
OR
THE BOY WHO WOULDN'T GROW UP.
A Play in Three Acts, by
J. M. BARRIE.

Peter Pan		Miss NINA BOUCICAULT
Mr. Darling		Mr. GERALD du MAURIER
Mrs. Darling		Miss DOROTHEA BAIRD
Wendy Moira Angela Darling		Miss HILDA TREVELYAN
John Napoleon Darling		Master GEORGE HERSEE
Michael Nicholas Darling		Miss WINIFRED GEOGHEGAN
Nana		Mr. ARTHUR LUPINO
Tinker Bell		Miss JANE WREN
Tootles, Nibs, Slightly, Curly, 1st Twin, 2nd Twin	(Members of Peter's Band)	Miss JOAN BURNETT, Miss CHRISTINE SILVER, Mr. A. W. BASKCOMB, Miss ALICE DUBARRY, Miss PAULINE CHASE, Miss PHYLLIS BEADON
Jas. Hook	(The Pirate Captain)	Mr. GERALD du MAURIER
Smee, Gentleman Starkey, Cookson, Cecco, Mullins, Jukes, Noodler	(Pirates)	Mr. GEORGE SHELTON, Mr. SYDNEY HARCOURT, Mr. CHARLES TREVOR, Mr. FREDERICK ANNERLEY, Mr. HUBERT WILLIS, Mr. JAMES ENGLISH, Mr. JOHN KELT
Great Big Little Panther, Tiger Lily	(Redskins)	Mr. PHILIP DARWIN, Miss MIRIAM NESBITT
Liza	(Author of the Play)	Miss ELA Q. MAY

Flyer for the opening of J. M. Barrie's play *Peter Pan* on December 27, 1904, at the Duke of York's Theatre in London.

during which the microscope had rudely replaced fairies with bacteria, vitality with calculation, youthful imagination with scientific law. Peter Pan became a tool of the opposition: a mystical boy in a miniature world who embodied an increasingly disappearing spirit of freedom.

I knew all this because when I was younger everyone called me Peter Pan. I was the juvenile one in my family, the late bloomer, the one who walked around barefoot without a shirt, stayed out late climbing in the forest, and always ate food with his fingers. When I was a boy my mother told me that Peter Pan was a fantasy. But as a professor specializing in evolutionary thought, I also knew about the axolotl.

Issue #43 of *MAD* magazine from 1958 ran a poem called "I Wandered Lonely as a Clod":

I wandered lonely as a clod,
Just picking up old rags and bottles,
When onward on my way I plod,
I saw a host of axolotls;
Beside the lake, beneath the trees,
A sight to make a man's blood freeze.

Some had handles, some were plain;
They came in blue, red pink, and green.
A few were orange in the main;
The damndest sight I've ever seen.
The females gave a sprightly glance;
The male ones all wore knee-length pants.

Now oft, when on the couch I lie,
The doctor asks me what I see.
They flash upon my inward eye
And make me laugh in fiendish glee.
I find my solace then in bottles,
And I forget them axolotls.

A man encounters axolotls on a walk and goes crazy. Who can blame him? Of all the creatures evolution has divined, this one may be the weirdest. Never growing old? Freezing time in its tracks? This was supposed to be the stuff of Victorian fantasy, not science.

The history of the Mexican axolotl, *Ambystoma mexicanum*, is bound up with European colonialism. An expedition sent to Mexico by the French government to explore the country's resources returned in 1864 to Paris, and among its treasures were thirty-four fishlike lizards, axolotls—a favored Aztec delicacy. They were summarily presented, together with three does and three small Chihuahuas, to

Isidore Geoffroy Saint-Hilaire, the president of the Zoological Society of Acclimatization. The son of Étienne Geoffroy Saint-Hilaire, Isidore was a zoologist who had coined the word *ethology*. Charged with improving agricultural production in France and its colonies by introducing new species of animals and plants, he and his society were afforded a garden, often referred to as the world's first modern zoo. It was there that the axolotls now settled into their new home.

The new arrivals were unlike anything anyone had seen before. Ranging in length from six to fifteen inches, their heads were wide and their eyes were lidless. As *MAD*'s clod would later remark, they came in different colors: brown with gold speckles and olive undertone, pale pink with black eyes, albino with gold eyes, gray with black eyes, and all black with no gold speckles at all. They had barely visible teeth, and their limbs were underdeveloped. There was a caudal fin extending down their spine to the vent. But most conspicuous of all were the external gill stalks, originating behind the head. There were three on each side, lined with extravagant filaments to better help them breathe oxygen underwater. It was as if a strange-looking flower had blossomed behind their ears, creating a dragon-like crest.

As it happens, fifty years earlier none other than Alexander von Humboldt had sent Georges Cuvier two dead axolotl specimens preserved in alcohol while on his journey to South America. These were meant to help settle a question. Cuvier was the founder of comparative anatomy, professor at the Natural History Museum in Paris, and the strongest man of science in Europe. What he wanted to know was whether Carl Linnaeus had been right to ordain two separate classes, the reptiles and amphibians. Linnaeus had claimed that amphibians could breathe both on land and underwater, but what counted for Cuvier was whether they simultaneously possessed air- and water-breathing organs. If they didn't, according to the Frenchman, they could not be afforded special intermediary status between reptiles and fish.

Linnaeus had found that all amphibians breathed with lungs as adults, having used gills during their early life as underwater creatures. But Cuvier knew from Humboldt that, although axolotls were plentiful in the secluded lakes of the Valley of Mexico, no adult air-breathing axolotls had ever been found. With the help of a museum colleague, the professor of ichthyology and herpetology Constant Duméril, Cuvier made a comparison with two similar-looking species, the olm (*Proteus anguinus*) from the karst caves of what would later be known as Slovenia, and the siren (*Siren lacertina*) from the coastal plains of North America. What Cuvier found was puzzling: While the olm and the siren were obviously adults, and therefore legitimate members of the class Amphibia, the axolotl was different. Based on its cartilaginous skeleton and poorly developed reproductive organs, Cuvier declared it to be the larva of a yet-to-be-discovered salamander.

That's where things stood fifty years later, when the live axolotls arrived in Paris in 1864. Intrigued by the mystery, the current professor of ichthyology and herpetology at the museum, Duméril's own son, Auguste, requested and was granted six of the lot, five male and one female, for his menagerie. In the fall of that year, his studies confirmed Cuvier's result: The axolotls were mere tadpoles. But then a curious thing happened, surprising Duméril entirely. Arriving at the menagerie one morning, he found the aquarium full of babies; the axolotls had reproduced! By 1866, the six pioneers had given birth to eight hundred! Seven months later a further surprise arrived: Eleven of the axolotls had lost their gills and become land-dwelling animals, full-grown *Ambystoma*, even as their parents remained tadpoles. How could that be? Had the parents reproduced before becoming adults? If so, their metamorphosis was extremely unusual, to say the least.

Haunted by the conundrum, Duméril died in 1870 with no solution in hand. But before he died, he generously sent out hundreds of axolotls to other researchers at universities, natural history museums, zoological gardens, even to curious amateurs. The rapid distribution of the Mexican walking fish, as it was becoming known, coincided with the growing popularity of aquariums, which both brought aquatic life into people's homes and played a crucial role in turning the experiment into a mainstay of laboratories, a new mode and locale of scientific research. Being the first exotic aquarium animal to be successfully bred in captivity, and therefore available at a low price, axolotls became all the rage. And for scientists they soon took a central role in a big argument about the ways of evolution: the one surrounding Haeckel's biogenetic law.

Haeckel had argued that the ontogeny of "higher" forms recapitulates the phylogeny of their "lower" ancestors. This necessarily compressed the earlier developmental stages of the ancestors into a shorter time frame, meaning that "higher" embryos developed faster than the lower ones. Also, as Haeckel's comparative drawings showed, recapitulation meant that you would see the adult features of ancestors appear in the juvenile stages of descendants. But the axolotl did the exact opposite. How could one explain tadpoles making babies? Obviously, they had developed the requisite sex organs, but in all other manner remained babies themselves.

Only a small number of the rapidly growing European population of salamanders ever morphed into air-breathing adults; the rest, aptly named perennibranchiates (forever-gilled), committed themselves to perennial youth. It was a phenomenon in search of a name, and in 1872 a man called Georg Seidlitz designated the retention of larval features in sexually mature adults *paedomorphosis*, and a man by the name of Julius Kollman called the same thing *neoteny* just eleven years later.* Haeckel, for his part, remained unimpressed. It was true that the timing of development could be tweaked, a phenomenon he coined *heterochrony*. But while the axolotl was "an exceedingly curious case," he scored it a persistent ancestral type with the help of some intellectual gymnastics. The theory was worth fudging. Darwin's reception in Catholic France had been hostile, but Germany was a different story. Here evolution was part and parcel of a grand narrative of social progress. A stab at the biogenetic law

* To be precise, paedomorphosis is the acceleration of reproductive development with constant somatic, or bodily, development; whereas neoteny is delayed somatic development with constant reproductive development. Sort of the opposite, but with the same result.

meant a stab at the future of humanity. A whole lot rested on the slippery shoulders of the axolotl.

The German doyen of biology, August Weismann, stepped in to pass judgment. Going blind, and not too dexterous, Weismann contracted Marie von Chauvin, an independent researcher renowned as a skilled animal specialist. Placing five axolotls in an aquarium with an inclined floor used to mimic the shore of a lake, she gradually began lowering the water level to induce the salamanders to climb onto land. Some weeks later, the experiment was crowned a success: All five subjects had morphed into air-breathing newts. Weismann now pronounced: The Mexican walking fish was an instance of retrogression, its transformation a return to a more ancient form that had disappeared courtesy of natural selection. Clearly, what had been responsible was the accidental move from Mexico to Paris, and now the unnatural experiment by Chauvin. But fear not, Weismann calmed his audience, "The contradiction is only apparent." The biogenetic law still stands.

When Weismann died in 1914, the world's evolutionists stood behind him. Whether they called it an atavism or a reversion, a truncation or an instance of developmental arrest, most scientists who had anything at all to say about the axolotl deemed it an exception, and an unimportant one, to Haeckel's biogenetic law. Unbeknownst to them, the days of the law were numbered, but not because of the axolotl. The impending marriage of evolution and genetics would lay the doctrine of the inheritance of acquired traits to rest, but it would do something even more dramatic. There was no reason why evolution had to march relentlessly in one direction; in the lottery of genetic mutations, whatever worked best was what survived. Since there seemed little grounds for believing that individual development and the evolution of lineages were related, ontogeny and phylogeny were promptly divorced and separated. The very necessity for progress was being called into question.

I leave Yaeli at the hospital for a routine checkup, and go to visit my parents down the hill. Turning the key to my childhood home, I think of the saying that the child is the father of the man, but how in reality children become fathers and mothers of their parents. My folks, like most of the folks of our friends, are getting old.

It's not just my parents who are getting older, it's the entire world population. According to the United Nations, the number of older people has tripled in the last fifty years in absolute terms, and will more than triple again in the next half century. Demographic predictions are notoriously slippery, but the ones we have today say that 183 out of 195 territories and countries are on course to reach below-replacement-level birth rates by the end of the century. Just visit untold European villages with hardly a child playing ball outside, or consider that the Japanese diaper market for the aged recently surpassed the one for toddlers. Walking up the creaky steps to my parents' room on the third floor of our family home, I worry about what their future holds: Alzheimer's, Parkinson's, organ failure, cancer. While the world tries to figure out what having fewer children will do to us, all of us face the challenge of trying to stay young while growing old.

In the years following Weismann's death, it became clear just how special axolotls are. Normalized for body size, their lifespan is comparable to ours. Still, females can produce eggs throughout their lives, and axolotls are highly resistant to cancer. More amazing still is their ability to regenerate. If a leg or a tail is torn off by a hungry fish, it will grow back. If an eye is poked out by a protruding branch, a perfectly new one will appear within weeks. Even a spinal cord, or brain or heart, can be forged again. If only we could all be reborn as axolotls.

All living things are capable of wound repair; it's a conserved trait in nature and makes good evolutionary sense. Most creatures lose this trait as they get older. Not axolotls. Just like us, their cartilaginous skeleton is replaced by bone as they age; their dermal layers thicken; their youthful locomotion attenuates somewhat, suggesting a slowing down of metabolism. But besides being able to regenerate whole organs and limbs, what makes them unique is that they retain their regenerative abilities almost until the very end.* Compared to us, axolotls never lose their magic. A given axolotl may be eaten by a bird, or drown in a storm, or suffer starvation. But it will never die because it fails to regenerate its brain, or a foot.

If we could figure out just how axolotls do it, we might find new ways to treat age-related decline in humans. But this scares some people. Undertaking the task of turning back time is immodest, they claim, and will produce consequences well beyond what our minds can comprehend. Others say: anything and everything for the sake of our parents. They have grounds for optimism. Using the axolotl as a model, scientists have figured out which proteins can substitute for nerve signaling and induce tissue growth, and which control the entry of cells into their cell cycle; they've tweaked the genetic regulation of immune response in ways that support regeneration (a reduced immune response being integral to the regeneration process); they've found that scar-free healing is based on collagen reorganization and fibroblasts; and they're looking carefully at the ways axolotls use epigenetics to limit the spontaneous development of tumors, since they almost never get cancer. Recently, the entire axolotl genome was sequenced, and it won't be long before we find the crucial genes that help newts stay eternally young.

Most surprising of all has been what scientists have learned about metamorphosis. There seems to be a connection between undergoing, or rather, not undergoing, metamorphosis, and being able to regenerate. Axolotls can be induced to metamorphose (we've known that ever since Marie von Chauvin), and when they do they can still regenerate (frogs cannot). But when axolotls metamorphose, regeneration starts to creak: Extra or missing digits appear, tissues don't

* Axolotls' regenerative capacities decline, just a bit, perhaps because they grow indefinitely, and larger size adversely affects the speed of wound closure and tissue restoration rates. Or it may be due to the thickening of skin with age, and its consequent loss of flexibility, making it more difficult to form a wound epithelium.

cohere, and death comes in their wake. Something about not developing fully allows axolotls to prosper. It's as if staving off adulthood helps saves their lives.

This is where Peter Pan comes in again. Because if axolotls are paedomorphic, or neotenic, so, in many ways, are we.* Our ancestors were hairy and had projecting jaws and low vaulted craniums. We on the other hand have round heads, large eyes, a flat face and small jaws, and all in all are pretty smooth. When you think about it, we look a lot like baby apes, and have many other juvenile traits: short arms, small teeth, reduced brow ridges. Most significantly, when we're born, our brains are only a quarter grown. These are all neotenic features, and, unlike chimpanzees and gorillas, we retain many of them through adolescence. Sexual selection may have had something to do with it: Both females and males of our species seem to prefer neotenic features in the other sex over fully adult ones. The evolution of monogamy and parental care for young might have resulted: As our fondness for cute little things clarifies, neotenic features elicit the compulsion to care for another, to nurture and protect. Darwin himself noticed that such predilections became entwined over evolutionary time with powerful emotional responses: We disarm when we encounter juveniles, becoming shielding and more tender. We even cuddle the young of other species, like lion and bear cubs and baby salamanders and kittens and mice. Like everything about our natures, this too seeps into our culture. As Stephen Jay Gould reminded us, Mickey Mouse was born in 1928 in his debut performance in *Steamboat Willie* with a long snout and small eyes. No wonder Disney artists gradually shortened his nose, enlarged his eyes, rounded his head, and made his legs seem smaller by lowering the waistline of his trousers. The people at Disney probably

* Since the levers controlling the timing of development can turn both forward and backward, we're also what's called *peramorphic*, meaning that some of our traits, rather than being retarded, sped up in evolution (the technical definition is "the extension of development past ancestral ontogeny"). Temporally speaking, we're mishmashes.

knew nothing about neoteny, but the more Mickey Mouse became baby-like, the more he was loved, and the more he was loved, the greater the revenue.

Today we know that neoteny is all about timing, and that this is true for development more generally.* In humans, the top end of the embryo begins to differentiate first, growing more rapidly than the end where the feet begin to take shape. That's why our babies are born with large heads relative to their medium-sized body and tiny legs and feet. As we grow older this anterior-posterior gradient is reversed: Our legs grow faster, our heads grow slower, and our eyes stop growing completely, appearing less prominent in our gradually elongating face. In effect we're becoming more apelike, though relative to apes who accelerate these changes much more powerfully, we remain childlike. Babyhood in other mammals is marked by clumsy movements and erratic behavior, which becomes more channeled and regimented with age. But extending our clumsy, erratic childhood well beyond that of any monkey or ape, our brains continue to grow into adolescence, and even into adulthood. By tweaking the timing of the onset of development, evolution produced in us a creature like the axolotl: We grow older but we never entirely grow old. Maybe that's why J. M. Barrie's creation has become such a classic. For better or for worse, we're all a bunch of Peter Pans.

Except that we do, alas, grow old. At a certain point, we slow down and weaken, and begin to become dependent again, just like when we were kids. Before we die, our life cycle includes embryogenesis, fetal development, childhood, adolescence, adulthood, midlife, mature adulthood, and senescence. Some scholars believe that the longer our behavior remains flexible after birth, the more education and learning become important; that flexibility and ability to learn from mistakes and develop is what makes us masters of the world.

---

* The Nobel laureate geneticist Barbara McClintock proclaimed in 1989: "If I could control the time of gene expression, I could cause a fertilized snail egg to develop into an elephant. Their biochemistries are not that different; it's simply a matter of timing."

This may be true, but the bigger picture is more sobering. Depending on circumstances, the arrow of development can point backward or forward. By the time we began to understand genetics, the mute axolotl had spoken: There are no "lower" ancestors reliably tucked away in the embryos of their more "highly evolved" descendants. Evolution is no march forward. Humans are not poised at the apex of creation. For many this may actually come as a relief, a happy liberation. But it's also enough to drive a believing person mad.

"Come over here," my mother whispers from her bed. She gives me her hand, my father curled up gently beside her. I look at them both, smaller than I've ever remembered. They seem vulnerable and tired. But they have a secret twinkle in their eyes, the wisdom of the ages. My beautiful mother takes my hand and kisses it. "You've seen us," she smiles. "We love you."

"Now go back to your wife."

9.

RIDDLE

Many before Haeckel had attempted to place man in nature. Linneaus put us in the order of primates that includes monkeys, sloths, and bats. We were apes, to be sure, like chimpanzees and orangutans, but to him we weren't one variety but four: the American—copper-colored, choleric, and regulated by custom; the Asiatic—sooty, melancholic, and governed by opinions; the African—black, phlegmatic, languid, and governed by caprice; and finally the European—fair, sanguine, muscular, and governed by laws. Johann Friedrich Blumenbach, sometimes referred to as the founder of racial classifications, to the contrary, said *Homo sapiens* were five—the Caucasian, the Mongolian, the Malayan, the American, and the Ethiopian, commenting on the large penis of the latter—whereas Cuvier in France spoke of just three "races"—the white, the yellow, and the black. However many varieties of man there were, these luminaries noted the lowly state of the blacks and the high standing of white people, hierarchies that were expressions of the favor of God.*

---

* There was, in fact, an exception: Inspired by the British parliamentary debates on slavery, the German anatomist Friedrich Tiedemann (1781–1861) compared brains and skulls of different human groups relative to body size and found no marked differences between the races.

And yet, Darwin, and Haeckel in his footsteps, knew full well that man's evolution was a natural process. Darwin in gentle words and Haeckel with invective both rejected the view that there had been different divine beginnings to the different races: All had one origin in a primitive group of ape-men, and climate and sexual caprice had brought about the physical differences among our kind. Physical differences between races were merely skin-deep matters, relevant to aesthetics. The true fortunes of *Homo sapiens* had for some time been forged on another plane entirely: Natural selection worked most strongly now on the mind, not on the body.

Once again it was Haeckel's linguist friend Schleicher who had helped him see this clearly. The patterns of language descent perfectly mirrored the pattern of human descent, but the connection was even deeper. Languages were material expressions of the mind having evolved to fit its environment. And so, as languages evolved, they modified the brain to increase its size as well as its complexity. Made permanent by heredity, this fixed the races of different environments at different cultural and intellectual attainment. There was and could not be a separation between mind and body. It was this Monistic perspective that to Haeckel showed most clearly why the European had climbed so high while the Hottentot stayed so low.

In stem-tree after stem-tree, he made order of the races. And just like Goethe before him with his plants, he traced them all back to the *Urmensch*. Naturally, at the very *Spitze** for him were the Germans and Greco-Romans, and alongside them the industrious, trade-savvy and book-learning Berbers and Jews. It was not as easy an exercise as it may seem. Unlike physical traits that were easily measured, qualities of mind were elusive; Haeckel stated clearly that his hierarchies were therefore hypotheses, and that as knowledge of different peoples grew they'd need to be redrawn. In successive editions of his book some were demoted while others became elevated: The American Indians featured high on the scale of culture, for example, but were overtaken

* German for "pinnacle."

The Munich artist Gabriel von Max's 1894 rendering of Haeckel's ape-man without speech. From Haeckel's book *Natürlische Schöpfungsgeschichte* in 1868, 10th edition, 1902.

by the Japanese when the emperor Meiji proposed a constitution based on the one in Deutschland.

Now fossils of early progenitors were being discovered in Europe and beyond. With the certainty that Ariadne's thread connected apelike ancestors with African and Asian humans, he predicted a "missing link," and named him *Pithecanthropus alalus*—"ape-man without speech." Some years later Eugène Dubois* traveled to Java to search for him, making a discovery that Haeckel held greater for anthropology than Röntgen's invisible rays were for physics. Dubois wrote Haeckel a letter thanking him for the inspiration, and called the missing link he'd predicted *Pithecanthropus erectus*. Today we call him *Homo erectus*.

---

* Eugène Dubois (1858–1940) was a Dutch paleoanthropologist and geologist, famous for discovering "Java Man" in 1891.

Above all else stood the miraculous three-fold parallel between ontogeny, phylogeny, and systematics, or what Haeckel thought of as the past, the future, and the present. He was sure that the development of a higher organism passing through the entire evolutionary history of its ancestors was the strongest evidence for descent theory yet. An exacting anatomist, not a soft-hearted dreamer, Haeckel thought of himself as always putting the facts before the rest. He saw clearly that if recapitulation was always complete, it would be an easy task to construct whole phylogenies on the basis of ontogeny, but this wasn't always the case. Responsible for this difficulty were the conditions of existence which led embryonic forms to be changed themselves, like the larvae of sea squirts that had developed streamlined tails to swim in pelagic waters. Though his biogenetic law assumed that new and progressive features in evolution would always be added at the terminal stage of development, it was true that the heart and the brain appear early in vertebrates, even though they don't exist in primitive forms of invertebrates. It was also true that embryos sometimes resembled not the adults of past ancestors, but rather the larvae of early archetypes. Sometimes they showed no resemblance to anything known to us at all. All this Haeckel knew. And so he invented the term *heterochrony* to denote that there might be changes in the timing of development of one organ or characteristic of an animal with respect to the rest; certain features of a species might be progressive whereas others might be regressed. There were myriad clocks in nature, as with the myriad stations of a man's heart, some looking forward, others shackled to the past.*

Still, his mind was gripped by recapitulation. How it elegantly solved the problem of the gaps in the fossil record, and by the beauty and truth of it on the grand scale of being. If life sometimes didn't

---

* Haeckel also invented the term *heterotopy* to denote a displacement not in time, but in place. His favorite example of this was how cells that differentiate into reproductive organs in modern organisms come from the mesoderm, whereas these organs must have arisen historically in one of the two primary layers, either the endoderm or exoderm, which came before it.

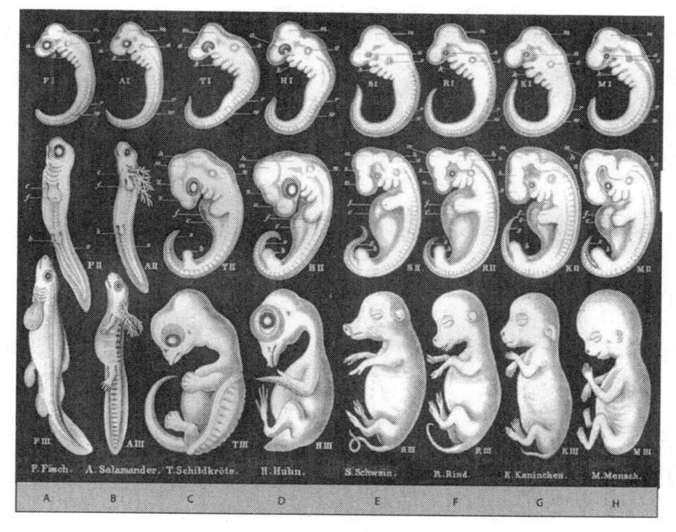

Haeckel's plate comparing the development of different vertebrate taxa (left to right: fish; salamander; turtle; chicken; pig; cow; rabbit; human). From his *The Evolution of Man* (1874).

move in a straight line, he willed, that was an exception.* He gave lectures in packed auditoria, using enlarged illustrations of developing embryos to convey the truth of his biogenetic law. In a book meant to reach even larger audiences, he showed a fish, a salamander, a turtle, a chicken, a pig, a calf, a rabbit, and a human—side by side at three stages of development. Earlier stages were almost identical in all the creatures, pointing to a simple common ancestor in the distant past. A human embryo began with gills like the fish that it came from, and retained a tail a little longer like the monkey swinging in a tree. There could be no better proof, he thought, of evolution going forward: The embryological record showed manifestly along with fossils and the tree of life that creatures were climbing higher and higher.

---

* Haeckel coined two new words to signify what he called "epitomized history," and "falsified history": *Palingenesis* described terminal additions that swept evolution forward, whereas *cenogenesis* described the addition of new stages in the midst of development.

At Jena all the while, Haeckel expanded the ambition of his program. He built a new home that he named Villa Medusa, created a new Zoological Institute down the hill, and petitioned government to include the study of descent theory in German classrooms. In this, to his great regret, he was opposed by his former teacher, Rudolf Virchow, whose eyes had been blinded from seeing Darwin's truth. Virchow the eminent scientist had become more of a politician, asserting not only that no evidence existed for evolution but also that descent theory may lead to chaos, and even communism. Haeckel thought he had it backward: The theory of descent taught that equality was a lie, nothing but a pipe dream. This didn't stop Virchow from frightening audiences with the example of the French Commune,* signaling to Bismarck that the same thing might happen in Germany. "May your shadow never be less," Darwin's "bulldog," Thomas Huxley, strengthened Haeckel in his colorful language from London, "and may all your enemies, unbelieving dogs who resist the Prophet of Evolution, be defiled by the sitting of jackasses upon their grandmother's graves." Haeckel laughed when he received Huxley's letter, but when Darwin died on April 12, 1881, it was like being struck by lightning. He was alone in his homeland, fighting powers of darkness. But this only strengthened his resolve. The teaching of religious myths from the Middle Ages in schools had to be replaced by a rationally based natural religion, for the betterment of human society. The deadliest enemy of reason was not malice, but ignorance.

---

* Following defeat in the Franco-Prussian War of 1870–1871, the local government in Paris refused to accept occupation and organized to inaugurate social reforms, including separation of religion and state and the cancellation of rents. Backed by the king, the conservative national government in Versailles sent its troops into Paris, leading to a civil battle in which tens of thousands of Communards, many of whom were socialists, anarchists, and republicans, were executed and jailed. Darwinism was associated in France with socialism and revolution, in no small measure thanks to the first translation into French of the *Origin of Species* by the outspoken, self-taught feminist and freethinker Clémence Royer.

There were many myths and men to combat, including the successor of his beloved teacher Müller, a physiologist named Emil du Bois-Reymond* in Berlin. Unlike Virchow, Bois-Reymond had seen the truth of Darwin, but had shied away from the consequences. In a famous speech to the Academy of Sciences in Berlin in 1880, he spoke of seven great riddles. Three of them—the origin of life, the apparently preordained orderly arrangement of nature, and the origin of speech—he found difficult, but solvable in principle. Three others—the nature of matter and force, the origin of motion, and the birth of consciousness—he thought transcendental and out of human reach. Finally, on the question of free will, Bois-Reymond remained agnostic, contributing in Haeckel's view to the reign of superstition over reason. He called himself a strict mechanist, but failed to see that Haeckel's monistic philosophy was concerned first and foremost with the problem of substance. Showing that there could be no mind without matter and no matter without mind, monism provided solutions to the three riddles the Berlin orator considered insoluble. Those he thought difficult, including the birth of consciousness, were in turn answered by Darwin's theory of descent. The seventh riddle was no riddle at all but based purely on dogma: Free will was an illusion invented by mystics and philosophers, and used by organized religion to inspire shame and guilt. Bois-Reymond preached *ignoramus et ignorabimus*—we are ignorant and shall remain ignorant, but Haeckel put this to the lie in his book *Die Welträthsel.* Translated into English as *The Riddle of the Universe*, it was the bestseller of the age, selling four hundred thousand copies before the Great War.

Haeckel began painting landscapes once more, and traveling, drawn to the richness of life's profusion. Maybe this was an escape from the assaults hurled at him from different quarters, university lecterns as much as church pews across the land. He was called a

---

* Emil du Bois-Reymond (1818–1896) was a renowned physiologist famous for developing experimental electrophysiology, and being the codiscoverer of the nerve action potential.

fake, a cheat, a mere popularizer. Men of God accused him of materialism, men of science, of *telos*. The preacher of the Hampstead Congregationalist Church in England warned that he would steer the ship of humanity to primitive barbarism, sun worship, Mohammedanism, and self-love. Even *The New York Times* published an article in response to *The Riddle* gently titled: "Haeckel Kills the Soul." His best friend Carl Gegenbaur broke off their friendship, unable to withstand the pressure. But despite the obloquy, Haeckel remained true to himself. He did not wish, like his countryman Nietzsche, to replace Christian morality with a new Super Man beyond good and evil. He did not see in Darwin the death of God and loss of meaning. Instead, he perceived that altruism was a natural gift: The roots of the Golden Rule were to be found in even simpler animals, explained by self-preservation and social instinct. Haeckel's conscience was clear. From earliest youth he had studied just one book, the *Book of Nature*. It was through that book that he came to know the only true god, his god, the god of Goethe, and Spinoza.

Haeckel thought he saw clearly: Egoism was the key to self-preservation of the individual, and altruism the key to preservation of the species. Man's most urgent task was to strike a harmony between the two, notwithstanding the precept "If any man will take away thy coat, let him have thy cloak also," peddled by fools. Were not countless conveniences of technical science in industry and agriculture, commerce and hygiene, found by religionists to be injurious to man's soul, along with the pleasure that comes from painting, music, and sculpture? Did not Christianity sacrifice the ephemeral body for the immortal soul, putting no emphasis on corporeal exercise, including swimming and gymnastics? Was it not Paul who declared it better not to marry, and "good for a man not to touch a woman"? If humanity followed such pernicious advice, it would be killed off within two generations. Not only was the Church responsible for the death of over ten million persecuted heretics and victims of religious wars, it was the source of tenfold more misery cast upon those whose higher mental life was extinguished, whose conscience was tortured, whose family life was destroyed. Christianity, Haeckel preached, no less, is the enemy of civilization.

Monism was the foundation for a new theory of man and morality, but went far beyond mere fallible human beings. Those faithful to a personal God had extricated man from nature in order to serve Him, engendering a contempt for other forms of life, and other forms of matter. In reality evolution was an endless refinement of structure. A vast, all-embracing metamorphosis infused the entire universe: It could be traced in the geological history of the Earth, in the evolution of its living creatures and peoples, but also in glowing nebulae billions of miles away, embryos of stars and galaxies. Some of these harbored life, undoubtedly, although in shapes we can only imagine. While the embryo of a new world is being formed from a nebula in one corner of the universe, another has already condensed into a rotating sphere of liquid fire elsewhere, a third already cast off rings at its equator, a fourth become a vast sun whose planets have formed a secondary retinue of moons. Like the ancients, Haeckel saw now that all is eternally changing, all is in flux from the very start. But all is also a unity. Nietzsche was much too dour. He failed to see that indivisible into the organic and inorganic, into mind and body, God-Nature was an unfolding work of art.

More and more, Haeckel grasped that it was crucial that the people become educated; this was the nineteenth century, and yet the world persisted in believing medieval cant. Free will, resurrection of the dead, a personal God, eternal life in hell or heaven: All were rendered fictions by the intellectual triumphs of the times. The greatest of these was the law of substance, at the very core of the philosophy of monism. It encompassed two arms, the first being the law of the conservation of matter, stating that the sum of matter that fills infinite space is unchangeable. The second was the law of the conservation of energy, stating that the sum of force at work in infinite space is unchangeable. Both were inseparable, lacking meaning without the other, and attested to the unity of nature. At any moment, one force may be converted into another. Wedded to cell theory, the dynamic theory of heat, and finally Darwin's great theory of evolution, this upended everything, like a plow thrust into an anthill. And no one was happier than Haeckel to see those rudely awakened from their comforts scuttling about dazed and confused.

When the opportunity occasioned shortly before his retirement, Haeckel was eager to leave behind him a temple to his philosophy of nature. Adjacent to his institute and one hundred and fifty meters from his house, he built the Phyletic Museum, with a large tree of life painted on its façade flanked on each side by the words "ONTOGENIE" and "PHYLOGENIE." All could now see that man doesn't stand outside of or opposed to nature, but rather firmly within it, noble and ethereal. The museum would one day house glass menageries of butterflies, birds, and insects; a stuffed gorilla and the skeleton of an African elephant; beautiful models of human development; even a painted shoulder blade of a bowhead whale once owned by Goethe. Haeckel himself left the museum empty, but placed a bronze statue of a maiden holding the torch of truth in the foyer, for all to follow its light.

This would be his legacy: On one side, spiritual freedom and truth, reason and culture, evolution and progress, standing under the bright banner of science; on the other side, spiritual slavery, superstition, and falsehood, standing beneath the black flag of hierarchy. The battle between them was a human affair, but he never lost sight of the grander drama. No body is so small and no spirit so insignificant that it does not contain a part of the divine substance by which it is animated. From the lives of giant crystals to the souls of tiny radiolarian cells, the great is little, the little is great, everything after its kind.

In the shallow waters of Messina and the jungle depths of Ceylon, Haeckel bore witness to life eternally folding and unfolding, shaping and reshaping. This was the truth of things, from end to beginning: Substance preserves itself yet is ever changing. For Haeckel there remained therefore just one final riddle: When he dies, in some form or another, would he once more be united with his precious love,

Anna?

10.

LUCY

Metamorphosis is defined as "dramatic post-embryonic development." Notwithstanding monism, it applies to living things, here on planet Earth. And Kafka exempted, most of us don't think of ourselves as a metamorphosing species. But metamorphosis is a continuum, and its definition seems arbitrary. We may not molt, or grow wings, but considering our transformations into sexual adulthood, we too undergo dramatic change. As it turns out, our connection to metamorphosis in other creatures runs deeper than we imagine. Deep, that is, into the farther reaches of our African past.

We're sitting at breakfast, Yaeli's chair more than a foot away from the table, her belly's so large. Six months have passed, but the next three feel like an eternity. Eating our Cheerios in a cloud of morning exhaustion, everyone just sits there staring in silence.

Infanticide, I announce to Yaeli and the kids, out of thin blue air. In gorillas it's rampant. In fact it accounts for up to 33 percent of infant mortality, and in baboons it's even worse. It may seem harsh to us, but from an evolutionary perspective it makes perfect sense. Since the female only becomes receptive again once she's finished breastfeeding, getting rid of her baby is the fastest way to drive her into estrous. Females organize in a harem around a single silverback gorilla; chimps in larger mixed groups. It used to be believed that females are indiscriminate about their mates, and that males are solely preoccupied among themselves with gaining dominance. Today, we know that's not entirely true: Primates of all kinds and of both sexes have preferences, and they seem to engage in sex sometimes even when it doesn't lead directly to procreation. Still, reproduction and fertility stand above everything. Sex to beget babies remains the rule.

Not for us humans. We may not have sex with each other regardless of age, status, or gender, as bonobos do—to mediate social conflicts, from food sharing to reconciliation. But we hardly reserve it exclusively for reproduction, either, not by any means. Instead, we evolved in small, isolated populations with few sexual options, becoming picky, if randy: We seek out sexual partners bounteously but are highly discriminatory when it comes to procreation. Over time, for us humans, sex and reproduction became decoupled. Helping to strengthen commitment, the act took on another function, fostering genuine feelings of affection and love. In chimps, females sport swollen red behinds when in estrous, but our females developed concealed ovulation. Unable to determine when sex counted, males stuck around to ensure parenthood, and began investing in their young.

Yaeli and the kids look at me as if I've been hit on the head with a blunt object. Shaizee's almost six, Abie almost four now; both are old enough to recognize weird when they see it. And the last thing on Yaeli's mind is monkey sex, that's clear. She rolls her eyes at me: When I don't quite know what to do, I often start babbling about evolutionary theories. It's a silly but persistent vice, and I'm no genius at combatting my vices. So I smile at everyone stupidly. And just keep babbling on.

What was responsible for the gradual elimination of infanticide in humans? Most anthropologists say natural selection. Just as the arrival of bonobos on the southern bank of the Congo Basin river helped reduce violence, gradually selecting for more socially stable cooperative groups, so too would lower aggression and increased tolerance help improve survival in humans. Self-domestication was a group-level adaptation: As males fought less, the entire troupe would be more ecologically efficient, outcompeting less cooperative groups for resources, and siring more kin.

But not everyone agrees with this theory. For skeptics, it wasn't docile males who were responsible, but assertive females. Protecting themselves from larger males, mating with smaller ones, females eventually reduced the gap in size between the sexes from 25 to 35 percent to just 16 percent. By the time our ancestors walked the African plains three and a half million years ago, the physical difference between females and males had already shrunk appreciably. Sharp male canine teeth disappeared, and soon the rest would follow. Hominin males couldn't sexually coerce females anymore with great ease, nor would they engage in killing their children. Instead, they were kinder now, and more thoughtful.

As sex became decoupled from procreation, I continue, doing my best to catch Yaeli's eye, females began to enjoy it. One obvious reason was that they were selecting for social personality. But there was another reason, recently suggested by the Yale ornithologist Richard O. Prum: They were also choosing males with penises that made them happy. In relation to our cousins, we're really quite well endowed: Average penis length in male humans is six inches, compared to three in chimps and one and a half in gorillas. Not only that, ours sport a glans, a coronal ridge, and width, physical peculiarities that enhance female pleasure. Even the female orgasm, Prum believes, owes its *umph* to sexual, not natural, selection.

"What's your point?" Yaeli asks, wiping wayward milk off Abie's ears.

I am quick to admit that sexual selection went both ways. Male proclivities brought about breasts in females (we're the only mammal in which they're permanent, a result of aesthetics, not biology, since engorged teats are only obligatory while nursing). Curves, too, were born, and meaty lips and softer skin. None of these exist in the females of our ape kin. But by reifying the male gaze as an adaptation, a sexist bias has been woven into the fabric of evolutionary biology:

Breast size, waist-to-hip ratio, facial symmetry, and "femininity" have all been interpreted as signs of healthy genes—as if our male ancestors were making rational, productive choices. Few evolutionary psychologists, on the other hand, have gone on record claiming that penis size is an honest indicator of male quality. Perhaps these male scientists lack the courage of their own convictions.

In reality, the sexes have shaped each other in humans. Just look at body hair. Compared to our hairy ape cousins, our coats all but disappeared except in erogenous areas: armpits, the pubic region, on our heads, above our eyes. Why did this happen? Our ancestors learned to hunt by outrunning their prey; shedding hairy coats may well have been an adaptation to help us cool off more efficiently. Still, the fact that the patches of hair that do remain appear at about the same time in both sexes is a strong indication that they evolved as sexual cues.

What does any of this have to do with metamorphosis?

"What does it, *really*?" Yaeli asks impatiently, looking like she wants to kill me.

Puberty! I announce. After all, that's where the action is. Before puberty, little boys and girls are about the same when it comes to muscle mass and body fat. Their voices are indistinguishable, and so is the thickness of their skin. And yet within the span of several short years, usually beginning at eleven for girls and twelve for boys, young men will gain 150 percent more skeletal and muscle mass, and young women will gain twice as much body fat. Breasts will have grown (Shaizee smirks). The menstrual cycle will have started. Boys will suddenly notice facial hair, squirt semen ("Huh?" says Abie), and hear their voices crack. Puberty, though we don't count ourselves as metamorphic, is a form of dramatic post-embryonic development.

And it's not just our sexuality that's morphing. The maturing brain undergoes marked structural changes, too. Between the ages of twelve and twenty, there's a decrease in gray matter in the prefrontal cortex—a kind of "pruning" of unused connections. This results in a reduction in unnecessary "noise," associated with improvement in basic cognitive abilities and logical reasoning. Major changes to the density and distribution of dopamine receptors follow in the pathways connected to the limbic system, where emotions are processed and rewards and punishments experienced, and in the prefrontal cortex, often referred to as the brain's CEO. Since dopamine plays an important role in how pleasure is experienced, these changes influence sensation seeking. Fatty tissue called myelin wraps around neurons, speeding up signal transmission, especially between different brain regions, which is important for planning ahead, weighing risks and rewards, and making complicated decisions. Finally, the connections between the prefrontal cortex (the higher brain) and the limbic system (the more basic animal brain) strengthen—a crucial development for control over emotions. All these changes, it was once believed, were more or less complete by childhood. But recent research shows that they are the mark of puberty and early adulthood. Aristotle put his finger on it nearly two thousand five hundred years ago: "The youth are heated by Nature," he wrote, "as drunken men by wine."

MRI studies bear him out. Compared to the adolescent brain, the adult has many more resources—literally, more neural networks—for tasks requiring self-control. A further set of studies shows that when the brains of adolescents are scanned before the playing of a game in which attractive rewards—like piles of coins, or pictures of happy faces—are shown to them, the reward centers in their brains activate much more strongly than those of either children or adults. At the same time, there is comparatively very little activity in the prefrontal cortex, the region associated with decision-making and impulse control. No wonder the Supreme Court of the United States determined in 2012 that it's against the law to give life without parole to juvenile murderers. In a very real sense they are different creatures—like tadpoles and frogs, or caterpillars and butterflies—and shouldn't be judged by adult rules.

Juveniles are like Ferraris with fancy accelerators and no brakes, which might help explain why they are so much more prone to get into trouble. Unprotected sex, for one, or crazy kinds of driving. In the United States, three million juveniles—a number greater than the entire population of Chicago—are taken into custody every year, and over two million of these are arrested. Of those arrested, 80 percent under age eighteen will be arrested again within three years. National longitudinal studies of youth based on self-reporting show that "sensation-seeking" peaks at age seventeen, at which time "impulse control" is at a minimum. By the early twenties, the number of active offenders decreases by 50 percent, and by twenty-five nearly 85 percent of former delinquents completely desist from offending. Without the statistics we have now, or the fancy science, artists and thinkers have known this for years. Shakespeare's shepherd in *The Winter's Tale* wishes "there were no age between ten and three-and-twenty, or that youth would sleep out the rest." Closer to our day, Rousseau intoned that "as the roaring of the waves precedes the tempest, so the murmur of rising passions announces the tumultuous change. . . . Keep your hands upon the helm," he warned parents, "or all is lost."

I look over at Abie and he's nodding off, his face practically falling into his Cheerios.

"And do you know why adolescents are such a nightmare?" I ask Shaizee.

"Okay, why?"

"Hormones!"

Remember dopamine, and its role in brain development? Dopamine is a hormone that's released by the hypothalamus, but the hypothalamus is also the origin of hormonal pulses that in adolescent puberty tweak the levels of other hormones in the pituitary gland that surge and travel through the bloodstream in young girls to the ovaries, and in young boys to the testes. There they stimulate the synthesis and release of the sex steroid hormones, estrogen/progesterone and testosterone, which jump-start the reproductive system. Suddenly, girls get a monthly menstrual cycle, and boys discover that they make sperm. In both sexes they also bring about the secondary sexual characteristics: breasts, pubic hair, voice change, bone and muscle growth. The sex steroids feedback negatively on the hypothalamus and the pituitary gland to ensure that circulating levels don't get out of control.

Now imagine Lucy, I say as Shaizee wrinkles her nose at me, the three-million-year-old female specimen of *Australopithecus afarensis*, discovered in Ethiopia in 1974. Lucy is our ancestor, a member of a small troupe on the plains of Africa. She's short, just three feet seven inches tall, and weighs about sixty-four pounds. Skeletal analysis reveals that she could still hang and swing from branches. But Lucy is also endowed with an upright gait for walking. The ratio of her body parts is instructive: The length of her humerus to femur is about midway between modern humans and common chimps, indicating that either her arms were beginning to shorten, her legs to lengthen, or both. In some ways, she was made for walking even more than humans, since *Australopithecine* brains were still very small, and babies would therefore have had no difficulty squeezing out through the pelvis. Only later, when our ancestors' brains began to balloon, would childbirth become an ordeal, since we'd need to strike a balance in pelvis size, allowing females to both walk on two feet and (just barely) push out kids. We don't know whether Lucy was ever a mother. But she seems to have died naturally. Slightly worn molars indicate that she was a mature young adult, about twelve years old.

Lucy's canine teeth are what experts call more "spatulate" than those of other apes, meaning broad at the apex and tapered at the base. But what's most striking is that males of her kind had canines that were quite similar. Nor were their bodies much larger, showing only moderate difference in size between the sexes, just as in humans. Sexual selection in Lucy's troupe had already kicked in to deweaponize males, it seems, making them smaller and more friendly and less likely to attack. How did it happen? A good guess would be that female choice impacted the hormonal systems that kick in at puberty. Rather than choosing large, ferocious partners, inundated with testosterone, Lucy's own ancestors began to favor those whose testosterone levels were lower. As her descendants evolved into the forerunners of humans, the price for the transition may well have been the rocky road of adolescence. No other mammal, or ape, has such a long period of delayed maturation.

Laziness, mood swings, and impulsiveness can be as embarrassing for teenagers as the weird changes happening to their bodies, and infuriating to their teachers and parents. But maybe it's all been worth it. As humans gradually evolved from ape ancestors, there were other changes: the descent of the larynx for the development of rich spoken language, the reshaping of the opposable thumb for tool manipulation, even enhanced *T* and *B* immune cell function, possibly to allow us to live more healthily in ever-growing groups. But delayed maturation was a truly transformative adaptation. Extending the time for learning, it placed a premium on larger brains and sociality. Ultimately, as groups of hominids evolved to show smaller physical differences between the sexes and larger brains overall, less aggressive males began to take paternity more seriously, and a form of monogamy gradually set in. So did norms of reward and punishment. This led to the gradual evolution of the qualities we prize as evidence of our humanness—language, cooperation, and a moral sense.

And to think that what got it started was not divine dictates, or some universal law, but rather just selections, by Lucy and her friends, that amounted in evolutionary time to tweaks to our growth and development. We don't count ourselves as metamorphic, like jellyfish and axolotls, but a metamorphosis of a kind—one we call puberty that defines what we call adolescence—has allowed us to lower competition within our group. It also seems to have played a role in growing metaphorical wings for us—the ones that have taken us to the heights we call culture.

The sweep of it all is breathtaking. Haeckel, undoubtedly, would have been impressed. I think for a moment about his unending love for Anna. I remember how Yaeli surprised me, kissing me on our first date. I thought she was cute, but there were other cute girls out there. Infinitely smarter than me, Yaeli saw farther. I've never quite understood how she knew, at that very first date, what was still completely beyond me. "Have a good day," I chirp as I step out of the house with the kids wearing wool hats and carrying their lunchboxes. Turning to our bedroom, exhausted, Yaeli shuts the door behind me.

11.

STRAUSS

In August 1997, *Science* magazine published an article titled "Haeckel's Embryos: Fraud Rediscovered." An embryologist at St. George's Hospital Medical School in London named Michael Richardson had begun to wonder about Haeckel's famous drawings of developing embryos since they didn't quite square with his own research. He went on to perform a comparative study photographing embryos of the same age and species of fish, reptiles, birds, and mammals that Haeckel had, this time with special optical equipment. Haeckel exaggerated the similarities, Richardson found, and also failed to draw the embryos to scale. His conclusion: "It looks like it's turning out to be one of the most famous fakes in biology."

Richardson had not been the first to see this. In Haeckel's day his own peers had accused him of fraud, calling a number of drawings in *The History of Creation* "inventions." One notorious example were three woodcuts presented as unique renderings of a dog, a chicken, and a turtle at precisely the same embryonic stage, which were, in reality, precisely the same woodcut. Nor did it escape careful readers that in successive editions of his collection of public lectures, published in English under the title *The Evolution of Man*,

Haeckel massaged some of the stages of comparative embryo development to make them conform more closely to his biogenetic law. Echidnas from Australia and New Guinea, for example, develop their limbs much later than most other mammals, but Haeckel presented illustrations of echidna embryos omitting limb buds at early stages. Wilhelm His, the Swiss leader in the new science of developmental mechanics, was a particularly scathing critic. His had developed a microtome that could produce finely sliced sections of embryos, and claimed Haeckel was full of hot air. There were stages of development that he could not have seen with his eyes, simply because they didn't exist.

The years, and the battles, were taking a toll on Haeckel. His former student Ludwig Plate, now head of the Phyletic Museum that Haeckel had founded, summarily kicked him out without providing a convincing reason. Stunned, and moving his office to his house, Haeckel had a bad fall in 1911 and began walking with a stick. When the Great War broke out in 1914, twelve of his nephews and grandnephews marched to the front lines of battle, and within a year half would be either dead or wounded. His wife Agnes died in the spring of 1915, neurasthenic and depressed, leaving behind their neurasthenic and depressed daughter, the third offspring of their unhappy marriage. A granddaughter came to live with him now, and he was

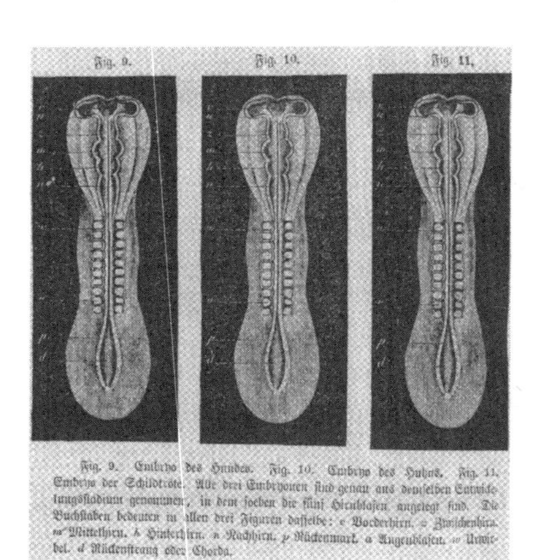

Woodcuts said to be of a dog, chicken, and turtle at the "sandal" stage (where the early embryo assumes the shape of a sole of a sandal) in *The History of Creation* (1868). All three were exactly the same woodcut.

growing very frail. Arrangements were made for Villa Medusa to be made into a museum—Haeckel-Haus—after his death.

But before all this, before the war, the fall, the dismissal from his own museum and posthumous humiliation, Haeckel was a star. He won medals from the Linnaean Society, the Royal Society of England, the Academy of Science of Turin, and was granted honorary degrees at the highest seats of learning: Edinburgh, Upsala, Geneva, Cambridge. He was the world's first popular science phenom, and sold more books than anyone. Revered by leading scientists like Ernst Mach, August Weismann, and Sigmund Freud, and revolutionaries like Lenin and Thomas Edison, he was inundated with thousands of letters from adoring fans from all over the world. One, from the dancer and choreographer Isadora Duncan in the spring of 1904, invited him to come to Bayreuth to attend Richard Wagner's opera, *Parsifal.* It was a story of a man's epic journey inward to discover the fullness of life and the quietude of the soul. As described by Haeckel's biographer, Cosima Wagner was in attendance and, not being a fan, was appalled to see the prophet of monism in the sanctuary of her late husband. Glad to stick it to her, Duncan promenaded glowingly during intermission in a Greek tunic, bare-legged and barefooted, hand in hand with Haeckel, whose white head towered above the multitude. When she gave birth to a son some years later she wrote to him: "This boy will be a Monist and who knows but some of your great and Beautiful Spirit may be in him." At the time, second perhaps only to Freud, Haeckel was the world's most famous living German.

On a par with Haeckel is a man whose head, tapering down at the chin, sits upon a small body, his stern gaze framed by a moustache and a receding hairline: the composer Richard Strauss. When he was young, Strauss's father, the principal horn player at the Court Opera in Munich, warned him to be suspicious of Wagner; he's too progressive, he told him, stick to Mendelssohn and Schumann and Bach. Strauss was drawn nonetheless to the festival at Bayreuth. Before long Wagner would become his greatest musical influence. Like Wagner, Strauss was enchanted by the Romantic philosophy of

Arthur Schopenhauer, according to which the observable world is the product of a blind and insatiable metaphysical will, not the received endowment of a classical order. And music is humankind's instrument of redemption.

In 1889, drawn by the lure of erotic freedom, mythmaking, and mysticism, he composes *Death and Transfiguration* (*Tod und Verklärung*), which cements his reputation. It describes an artist on his deathbed, dreaming of his innocent childhood, the struggles of manhood, the attainment of his goals, and, finally, the metamorphosis from the "infinite reaches of heaven" for which he yearns.

Strauss's rise is meteoric. A North American tour culminates at Carnegie Hall in New York, where he conducts the world premiere of a new symphony and performs a concert of lieder with his wife, the soprano Pauline de Ahna. Back at home, Gustav Mahler calls his

Richard Strauss.

opera *Salomé* "a live volcano, a subterranean fire." He is now a world celebrity. He and Pauline purchase a massive, white, three-story, thatched-roof villa in the small town of Garmisch, Bavaria.

Strauss tries to make sense of it all. Perhaps his leaping between the classical and the modern is what accounts for his success, the hearkening back to beloved traditions while simultaneously flirting with the modern idea of progress. But he too is transforming. Lately he's been moderating his harmonic language, forsaking dissonance for a lush melodic style. Rejecting the weight of Wagner's metaphysics, he is now composing "a music whose supreme charm is its ignorance of good and evil," as Nietzsche puts it, a "super-German" music colored by "the thousand tints of the setting of a moral world that men no longer understood."

Haeckel is listening to the tunes around him. Germany, the "machine nation," is leading the way in the chemical and steel industries, and in technologies of war. But a European obsession with biological degeneration and cultural decadence is particularly pronounced in the homeland, accompanied by popular anti-Semitism. Many agree with Nietzsche that faith in progress and rationalism has annihilated the mythic foundations of their ancient culture and produced the emptiness of contemporary life. Along with Wagner, painfully aware of the fragmentation of modern life and harboring a "hunger for wholeness," they call for a return to family and tribal values. Under these circumstances, a year before the outbreak of hostilities, Haeckel joins the French socialist and educational reformer Henriette Meyer to create the French-German Institute of Reconciliation to Prepare for Perpetual Peace between Our Two Countries. "Pacifism is a duty of humanity," he declares in their newly minted journal, *La Réconciliation*. But like many liberal German intellectuals, once the first shots of the Great War are fired in Belgium, Haeckel quickly changes his tune. Europe has grown jealous of Germany's economic and spiritual success, he claims, and Britain, in particular, is at fault. "It is with a bleeding heart," the great admirer of Darwin writes in his tract "England's Blood Guilt for World War, 1914,"

> and only because of the compulsion of my patriotic feeling that I, an eighty-year-old German citizen, have composed this complaint against brother England. For more than sixty years I have belonged to the group of those scholars who have held the mighty cultural work of Great Britain in the highest respect.

Haeckel is emotional. "A single finely educated German soldier," he writes, "possesses a higher intellectual and moral worth than a hundred of the raw, natural men whom England and France, Russia and Italy have brought to the front." Twenty million people would eventually lose their lives, and twenty-one million would be wounded. But true to his optimism about evolution pushing forward, Haeckel adds that the Great War could perhaps become a watershed to "open the eyes of men and in clear sunlight manifest the true worth of life on this earth."

When Germany loses her vast and bloody war, the German people rise in revolt: Kaiser Wilhelm II flees to The Netherlands, the Social Democrats proclaim the Weimar Republic, then sign the Treaty of Versailles. Soon the economy collapses. Historians will find in these circumstances—a humiliating defeat followed by a debilitating treaty, an untried government, and economic strife—the roots of the rise of Nazism. Haeckel, for his part, writes on Christmas Eve 1918 that the German nation, the pinnacle of evolution's stem-tree, now stands "almost without hope on the ruins of a ravaged culture." That its superior values and reason have been destroyed by the enemy comes to him as an utter shock. It can only be the case that alongside mankind's slowly evolving ideals of empathy and rationalism, a darker aggression and greed lurk within. If anything can lead to a better future, it will be a solid

A Couch's spadefoot toad (*Scaphiopus couchii*), looking, as Orwell imagined, "like a strict Anglo-Catholic towards the end of Lent." Credit: Gary Stolz.

A Maria Sibylla Merian painting from Surinam. Branch of unidentified tree with Menelaus Blue Morpho butterfly, 1702–3, Royal Collection Trust.

Above: Maria painting the creatures she encountered in the jungles of Surinam. Notice the giant tarantula eating a hummingbird. Branch of Guava tree with Army Ants, Pink-Toe Tarantulas, Huntsman Spiders, and Ruby-Topaz Hummingbird, 1702–3, Royal Collection Trust.

Left: A cicada having just emerged from its husk, known as an exuvia. The Chinese use the phrase "to shed the golden cicada" when they speak of fooling one's enemy by using a decoy. Credit: iStock by Getty Images.

A large blue butterfly (*Phengaris arion*) looking like a natural jewel.
Copyright © 2025 by Rickinghamphotography.co.uk.

An ocherous impcster larva of a large blue among red ants and their larvae.
Credit: T. Komatsu.

Above: Portrait of the reigning Empress of Russia Catherine the Great with her blue sash, Vigilius Eriksen, c. 1749. Courtesy of Rijksmuseum.

Left: Apollo and Daphne, after Ovid's *Metamorphoses*, Gian Lorenzo Bernini, 1625. The sculpture is housed in Galleria Borghesa in Rome.

Above: Watercolor of a pineapple with the life cycle of a Dido longwing butterfly depicted, plate 2 in Merian's *Metamorphosis Insectorum Surinamensum*.

Right: Portrait of Maria Sibylla Merian at thirty-two by her stepfather Jacob Marrel, 1679 (Kunstmuseum, Basel).

Mayflies emerging in the thousands from a European river. The swarm is so dense, it can be detected by Doppler radar. Credit: Alamy.

The immortal jellyfish (*Turritopsis dohrnii*) © Takashi Murai/The New York Times Syndicate/Redux.

Some years after he saw and named *Mitrocoma annae*, Haeckel discovered a second, and to his eyes more beautiful medusae, which he named *Desmonema annasethe*, the *Discomedusa*, rendered here in his book *Art Forms in Nature*, 1899–1904.

Above: *Goethe in the Roman Campagna* by Johann Tischbein (1786–7), the most famous portrait in Germany.

Left: Axolotl (*Ambystoma mexicanum*), or as *MAD* magazine called it, "the damndest sight I'd ever seen." Credit: Tim Flach.

Illustration of the European eel, *Anguila anguila*, from the 1879 book *British Fresh Water Fishes* by W. Houghton and A. F. Lydon.

The author, Yaeli, Abie, and Shaizee in the Red Sea, six months into the pregnancy, not knowing what the future holds. Credit: Oren Harman.

Different sea squirt species. Notice the structures that look like teeth, vestiges of chordate origins. Credit: Ofer Ben-Zvi.

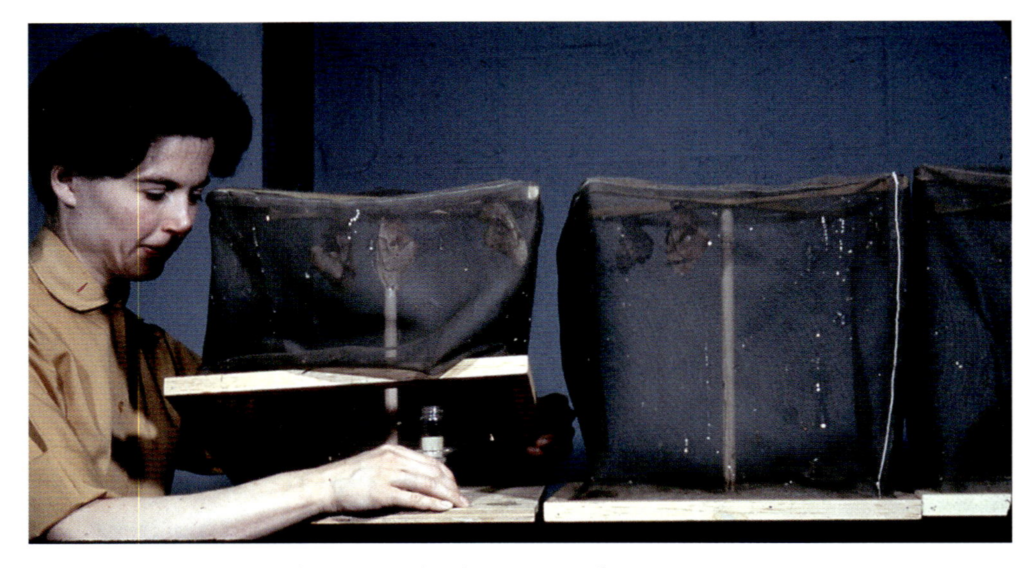

Lynn working with silk moths at Harvard in the 1960s. Credit: Jim Truman.

Dogs in a market listen to Rumi, who praises their understanding and attention. Credit: The Morgan Library & Museum, New York.

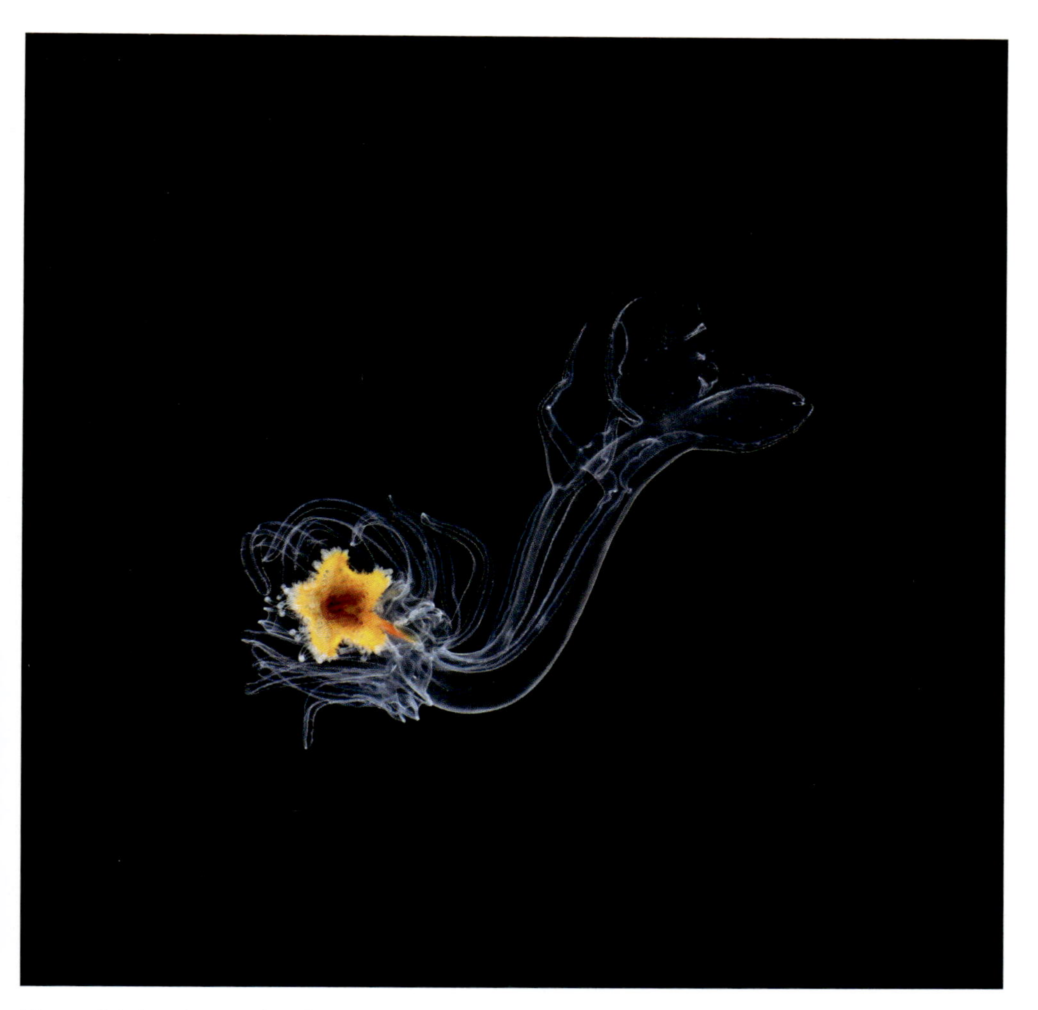

The starfish (*Luidia sarsii*) with the minuscule colorful adult growing from within the translucent larva, moments before they enact a truly immaculate conception. Credit: Richard KirbyRK_.

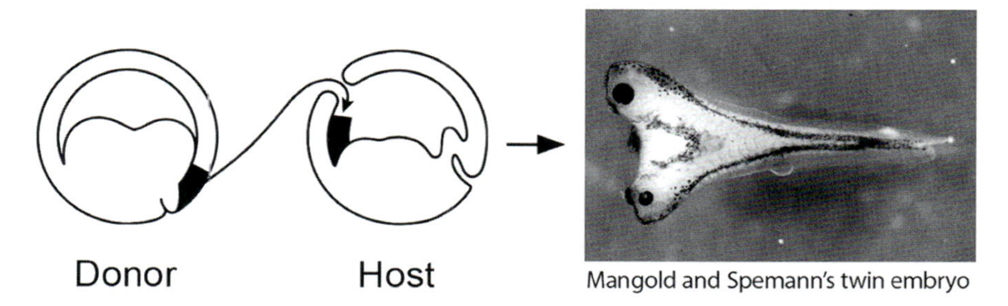

A modern reenactment of Hilde Mangold's experiment: twinned Xenopus embryo, with two separate notochords, resulting from an organizer graft. Credit: Dr. Edward M. De Robertis, UCLA.

Left: The tobacco hornworm, *Manduca sexta*, as rendered by the British entomologist and illustrator John Curtis, 1828.

Below: Lynn and Jim in their home on Friday Harbor in 2023, with Maria Sibylla Merian's 1705 Suriname book on the table, the "pièce de résistance" of their library. Photo by the author.

The strange-looking mouth and suckers of a lamprey, an ancient link between humans and the sea. Credit: copyright © blickwinkel / Alamy Stock Photo.

Progressive changes in the same pair of neurons through the metamorphosis of a fly brain, what Jim calls "a numbers game." Credit: Jim Truman.

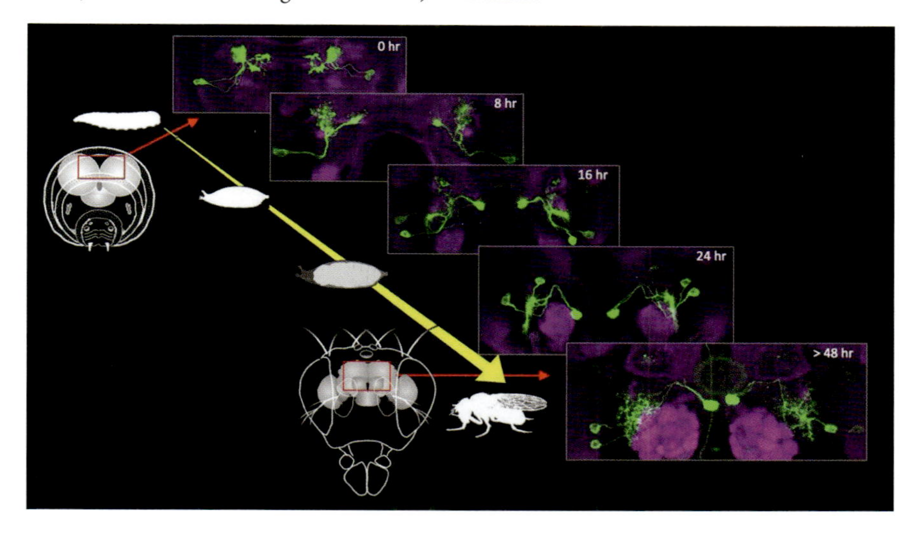

The wonderous diversity of insect body plans. Credit: Shutterstock.

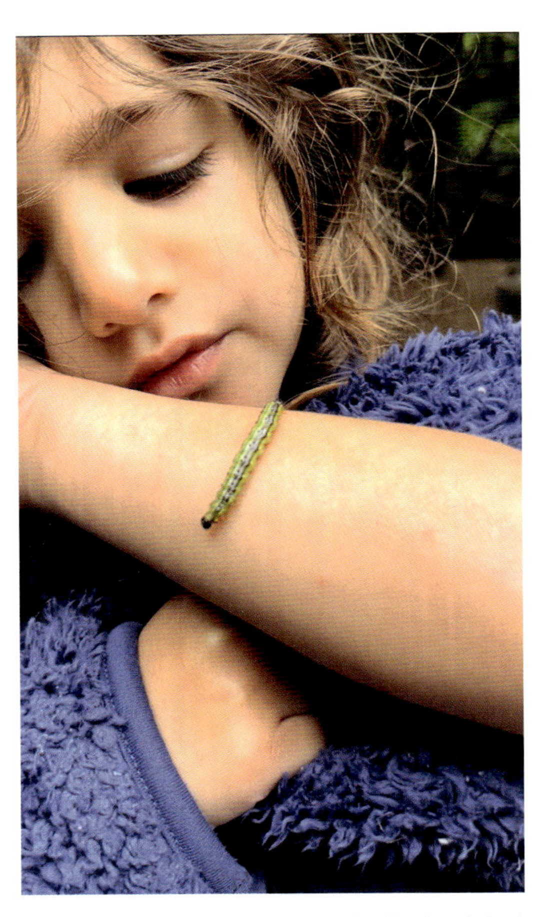

Sol, age 4, with a friend (photographed by her father).

school education for children, based on the monistic evolutionary doctrine.

Suffering from a weak heart, Haeckel expresses his "desired entrance to eternal rest." After a series of falls and fainting spells, on the morning of August 9, 1919, Ernst Haeckel dies in his sleep.

Like Haeckel, Strauss is a man torn between a rational and artistic disposition. In the aftermath of the Great War, he persuades himself that art and politics can be kept separate since one is ethereal and immortal whereas the other is transitory and corrupt. When the Nazis come to power, he decides to play along with them. Hitler is an admirer of Wagner and, ever since *Salomé*, of his music, too. He'll ingratiate himself with the yokel, he decides, the better to promote high German culture and art.

When the Minister of Propaganda Joseph Goebbels appoints him president of the newly founded Reich Chamber of Music, Strauss accepts the honor, writing to his family, "I made music under the Kaiser, and under Ebert.* I'll survive under this one as well." He doesn't join the Party, or condone any of its creed, but accepts the invitation to substitute for the Jewish conductor Bruno Walter when he's forcibly removed as conductor of the Berlin Philharmonic. He also replaces Arturo Toscanini when Toscanini resigns as director of the Bayreuth Festival in protest against the Nazis. That same year, 1933, Strauss gives as a wedding present the manuscript score of his new opera *Arabella* to Hermann Göring, who has just created the Gestapo, and he dedicates an orchestral song to Goebbels (a man he describes in private as a pipsqueak). Meanwhile, Jewish businesses are being shut all over town, performances of Mendelssohn and Mahler banned, the books and papers of Freud and Einstein burned in the street.

Strauss isn't worried. Art is transcendent, he's convinced, impervious to the follies of nations, parties, and power-hungry men. He

* Friedrich Ebert (1871–1925) was a German politician from the Social Democratic Party, and president of Germany from 1919 until his death.

refuses to see that art can be turned to political ends. His "Olympic Hymn" is performed at the 1936 Berlin Summer Olympics; his carefully crafted chromatic harmonies and orchestral tones provide an exonerating soundtrack for the Third Reich. Then things get more personal. Strauss and Pauline's only son, Franz, is married to a Jewish woman. When the Nazis round up the Jews of Garmisch-Partenkirchen, the village where Strauss and his family live, his own two Jewish grandchildren are among them. In a perverse reflection of their mixed racial heritage, the children are both spat on and made to spit on others. Strauss's influence is waning. Exasperated by his insistence on working with the Jewish novelist and librettist Stefan Zweig, Goebbels strips him of his directorship of the Reich Chamber of Music. In March 1944 Strauss travels with his chauffeur to Theresienstadt concentration camp in an effort to save his daughter-in-law's grandmother and twenty-five other relatives, but he is laughed away by the sentry at the gates. The grandmother is murdered with all the rest.

By now all the great opera houses have been destroyed: Munich, Dresden, Leipzig, Vienna. Holed up with Pauline in their country villa, in the summer of 1944, Strauss turns inward. He was sure he could ride the beast, but instead the beast has ridden him. In a notebook he pens the old lines from Goethe: "No one can know himself / Separate himself from his inner being," and "What happens in the world / No one actually understands." He begins to compose music to the words, but abandons the task, creating a wordless composition for twenty-three solo string instruments instead. On the last page of the score, a quote appears from the funeral march in Beethoven's symphony, *Eroica*, and below it the cryptic words: "IN MEMORIAM!" Years later the music critic and historian Jeremy Eichler will write that *Metamorphosen* is "an elegy to German culture, a death mask in sound." It is also one of Strauss's most moving creations, a "rueful meditation on the opacity of the self." Strauss dies quietly on September 8, 1949, a man overtaken by history.

In years to come, Strauss will be judged by some but loved by many. Haeckel will be treated much more harshly. Monism inspired fascism throughout Europe, the historian Daniel Gasman will write in the 1970s, and the paleontologist and essayist Stephen Jay Gould agrees: Haeckel's irrational mysticism and championing of biological laws of progress, his belief in Arian superiority and in the extermination of "lower" races in the struggle for existence, "all contributed to the rise of Nazism."

Is Haeckel the true intellectual father of Nazism? In the nineteenth century, racism was an affliction from which few were immune. Most believed, like Haeckel and Darwin, that "savage" peoples were being replaced in evolution by "civilized" Europeans and that this was how things should be. Yes, certain Nazis took to Haeckel's worldview. But Haeckel had ranked Jews next to Germans on the highest branches of his evolutionary progression. And his monism was officially repudiated by the Nazi Party. Yes, Haeckel was a complicated man, with a complicated legacy. Despite his racism, it makes sense that the heavily Christian Nazis rejected the teachings claimed for the left by Lenin and the Marxist leader Karl Kautsky.

In recent years, Haeckel has been enjoying something of a reassessment. Perhaps that's because the core of his scientific worldview still stands: There is a continuity between the living and nonliving; natural law does need to account for all physical phenomena; evolution by natural selection is a fact. Haeckel's insistence that there were other mechanisms besides natural selection at work in evolution, like inheritance of acquired traits or direct action of the environment, is increasingly appreciated by modern biologists. Many have tried to divorce the two, but Haeckel and Darwin usually agreed on matters of substance. Where they differed most was not in content but in style: Darwin was soft, gentle, appeasing, Haeckel was aggressive and commandeering. Darwin was self-doubting, Haeckel cocksure. Darwin shied away from the implications of evolution; Haeckel turned them into a religion. Had his brand of evolutionism not been so stinging to people of faith, they might have found less to dislike about the theory of natural selection. Darwin's conciliatory tone won over mainline protestants in the United States and England, but Haeckel got under people's skin. It's probably at least partly true that by some circuitous route we owe to him the demand by certain evangelicals that evolution and creationism be given equal time in schools.*

We no longer believe in progress, as Haeckel understood it, or in racial hierarchies; yet his synthesis of form and function remains relevant to biologists. For many years, the biogenetic law was ridiculed, but it too is making something of a comeback. Not as a law, per se, but as a fruitful concept. Today we know that genes control the rate and direction of embryonic development, and that genes can mutate. Sometimes mutations add new stages at the end of embryonic development, sometimes in the middle or beginning, and sometimes they erase stages entirely. Developing embryos aren't telescopes of a perfect evolutionary ascent from "lower" creatures to "higher" ones, as Haeckel himself conceded.** But embryos do reflect the course of evolution, even if that course is Rube Goldbergesque.

Haeckel's own evolution, as a man, included one final development, a secret that came to light some years after his death. In 1898, at the age of sixty-five, Haeckel had fallen in love again with a young woman named Frida von Uslar-Gleichen. In nine hundred letters exchanged

* After the *Science* article accusing him of fraud appeared in 1997, creationists were quick to attack evolution based on the association made by historians between Haeckel and the Nazis.

** Haeckel's full description of the biogenetic law was: "The rapid and brief ontogeny is a condensed synopsis of the long and slow history of the stem (ontogeny): this synopsis is the more faithful and complete in proportion as palingenesis has been preserved by heredity and cenogenesis has not been introduced by adaptation."

over six years, a story of growing passion unfolds, including a romantic visit to Goethe's house in Weimar, their first time holding hands, then furtive kisses, followed by passionate trysts in off-the-beaten-track hotels. Haeckel wrote to Frida that she was the "bride of [his] soul," his "best, truest wife," the incarnation of Anna. (That she was born in 1864, the very same year Anna died, seemed to him no coincidence.) Frida, for her part, called Haeckel "my silver bunny" and "my sacred one," and begged him to reveal their affair to his wife. Haeckel demurred, writing to her that a superstitious church and bourgeois morals wouldn't allow for their heavenly union, nor, on account of her weak heart, would his wife survive the news. Deeply disappointed, Frida's heart weakened, too, and she was sent to convalesce in a sanitorium. The couple wrote pining letters in which they considered a suicide pact, and Haeckel even sent her morphine. Finally, in mid-December 1903, by way of an "accident" reported to Haeckel by her sister, Frida climbed up the staircase in her home in the middle of the night, collapsed, and died.

Soon after Frida's untimely death, Haeckel published a book called *Art Forms in Nature*. Filled with over one hundred detailed, multicolored illustrations of animals and sea creatures, it would have a profound influence on art nouveau in architecture, painting, and design. In one of the illustrations in his book, Haeckel depicted a medusa that he christened *Rhopilema frida*, in memory of his "artistic friend of nature." Once again he'd found in the medusa a repository of immortal love.

But there remained one final riddle: the substance at the core of all experience. How could it manifest both as matter and spirit, and why did it constantly morph? Haeckel went to his grave knowing that what scientists call Nature or the universe, what philosophers call Substance or the Cosmos and believers the Creator or God, was a complete and utter mystery. The deeper man probed, the more enigmatic it became. The investigation of this greatest of all riddles was best left, he thought, to the "fruitless brooding" of the metaphysician. Words could never describe it. Like Strauss in *Metamorphosen*,

The medusa *Rhopilema frida*. From Haeckel's *Art Forms in Nature*, 1904.

Haeckel concluded that the only thing we can do in the face of inexorable change is to bow our heads in resignation. Maybe music was how to express this, or some other form of art. As Goethe wrote:

By eternal laws
Of iron ruled
Must all fulfil
The cycle of
Their destiny

12.

EEL

For centuries, the origin of the European eel baffled the greatest minds. No one had ever seen an eel have sex, or even managed to locate its genitalia. Eels "proceed neither from pair, nor from an egg," Aristotle wrote in his *Historia Animalium*, instead they are born of the "earth's guts." Pliny the Elder thought they reproduced by rubbing themselves against rocks. According to other theories, they emerged from the gills of fish, from morning dew, from unknown "electrical disturbances." There was even a bishop who reported to the Royal Society that, their eggs having been incubated by the sun, eels were born on thatched roofs.

Leeuwenhoek claimed that the snakelike fish produced live young, but the baby eels he observed swimming in what he presumed was their mother's uterus turned out to be parasitic worms hunching in the bladder. Even Linnaeus wrongly deduced that eels must be viviparous: Dissecting an imposter fish named the eelpout, the man who was to order nature bungled things entirely. But perhaps the weirdest theory came from a factory engineer from Scotland. "The reader may at once be informed that . . . the progenitor of the silver eel is a small beetle," David Cairncross announced in a book that purported to sixty years of careful experiments. Alas, he too was just observing parasites.

Humans and eels have a colorful history. It is reported that six thousand eels were served at a banquet celebrating one of Julius Caesar's victories—in a recommended sauce of "dry mint, rue berries, hard yolks, pepper, lovage, mead, vinegar broth and oil"—and it was the Italians, fittingly, who became obsessed with "the eel question." In the wetlands of the mighty river Po, they built the world's greatest fisheries; at the height of the season, they hauled in hundreds of tons of eel per night. In 1707, one professor in Padua thought he detected ovaries but had to concede upon further examination that these "ovaries" were merely a swim bladder, badly diseased. Deeming it "a matter of extreme importance to find the true ovaries of the eel," a reward was announced for the first fisherman to produce a specimen with roe in it. Poor professors, how little they know of fishermen: The reward was duly canceled when it turned out that eels were being packed with the eggs of other draw.

The ovaries of eels were finally detected in 1777, but even then, the eel's testicles were nowhere to be found. For a hundred years scientists searched in vain. That's when Sigmund Freud arrived to the field zoological station in Trieste, aged twenty and intent. It was his first research job, and he would find those testicles if his scientific career depended on it. Every day, from eight till five, four straight hot summer weeks long, he cut open eels to find their private parts. There was no other way, he wrote to a friend, "seeing that eels keep no diaries." But Freud's eels all seemed to belong to the fairer sex. In his first-ever scientific publication, "Observations on the Form and the Fine Structure of Looped Organs of the Eel, Organs Considered as Testes," he admitted defeat: Eels must have balls, but he had failed to discover them. As the writer Lucy Cooke put it, "He went on to probe less slippery subjects, like the human psyche, with significantly more success."

Finally, toward the turn of the twentieth century, a male eel bursting with sperm was caught off the coast of Sicily and proved

Aristotle wrong: Eels must come from sex. The man who found the testes was the Italian Giovanni Grassi. But Grassi had already made no less momentous a discovery: For years fishermen had noticed tiny, transparent, leaflike creatures with bulbous black eyes and strange-looking buck teeth washing up on the shores in great numbers. "Thin-head, short-nosed," they were called, or in Latin *Leptocephalus brevirostris*. No one knew what life-form this was, but Grassi had a hunch. Counting their vertebrae, he found 115, and looked for a match elsewhere in the ocean. When he found it in the European freshwater eel, the picture suddenly made sense to him: The tiny leaves, "slivers of life," were in fact the larvae of the adult. *Anguila anguila* had a life cycle. The big question was, Where is it born?

Grassi proclaimed that the baby eel larvae hatched from eggs in the Strait of Messina, but failed to notice that the thin-heads he captured there were already almost three inches long. And so it was left to a pathologically ambitious Dane to detect their true origin. The oceanographer Johannes Schmidt began his career aboard the research vessel *Thor*, studying the mating habits of cod and herring. But in 1903, when the *Thor* was sailing near the Faroe Islands, the boat's fine-meshed trawl caught a tiny larva. It was the European eel, 2,500 miles north of Italy! Unless it had gotten terribly lost, the origin of this minute eel could not be the Straits of Messina. After twenty further years searching for the source, Schmidt found what he was looking for: a thin-head short-nose just 5 millimeters long. The larva was so small that Schmidt was convinced it could not have originated anywhere else. "Here lie the breeding grounds of the eel," he wrote, elated but exhausted. It was April 12, 1921, in the warm and stunningly clear waters between Puerto Rico and Bermuda, otherwise known as the Sargasso Sea.

After centuries of befuddlement, it didn't take long now to discover: As a tiny "sliver of life," the eel drifts for two or even three years in the currents of the brutal North Atlantic Ocean. In a journey 6,000 kilometers long, fewer than one in five hundred will reach the shores of Europe or North Africa. By the time of its happy arrival at the mouth of a Swedish or Moroccan river, it has grown into a "glass eel," three inches long, with a proper head, and eyes, and tail, and soon even a new name—elver. It darkens to avoid predators, swimming against the tide, crossing land ditches, climbing vertical surfaces, doing its best to avoid menacing birds, otters, and humans. When it finds a good den at the river bed, it clings to it, never venturing far from home.

Many years pass, sometimes more than fifty, and *Anguila anguila* is a massive yellow eel now, three feet long. Its jaws are strong, it is serpentine and muscular, a solitary and fearsome hunter. After yet another cold winter spent lying motionless in its lair, it awakens and begins to head down river. Finally, it is sexually mature. The European eel is ready for its long journey home.

Now it undergoes its fourth, and final, metamorphosis. The yellowish-brown tinge turns silver gray with dotted black stripes running down its sides. When autumn comes, the eel slips back into the Atlantic. Fins grow longer and stronger to propel it, eyes grow larger and blue to allow vision into the depths. All the eel's chemistry has changed to accommodate salt water. Its digestive system shuts down, its energy is stored in fat reserves. The eel will not eat again. Like a fearless arrow, a kilometer beneath the surface, it swims thirty miles every day. There are sharks and whales down there, and trawling, eel-meat-loving humans. But filled with roe now, *Anguila anguila* is single-minded. It has a cycle to complete before it expires. Its destination: the Sargasso Sea.

As the days pass, Yaeli is getting more and more tired. For her, the destination seems far off, still. She stares at me beneath the covers one night, her soft eyes asking, *Where are we going?* I'm wondering the same. The music of four has finally been coming into harmony, but what will be the sound of five? As Yaeli's eyes shut, I imagine millions of eggs falling in aqua waters from their exhausted eel mother. I think of the immortal jellyfish, of Haeckel and Goethe, of Lucy and the axolotl. Maybe the cycle of life is a forward lunge toward beginnings. Maybe, as Strauss composed, those beginnings are actually forms of endings. I'm confused, somewhat somber, but one thing is for sure: The birth of our baby will be an upheaval.

All of our senses of self are about to change.

PART THREE

WHAT IS THE SELF?

1.

MOUNT MONADNOCK

Standing proudly at 3,165 feet, Mount Monadnock rises above the red spruce plain of southern New Hampshire, bare boulders of 400-million-year-old schist and quartzite strewn on its naked summit. In the indigenous Abenaki language, its name means "mountain that stands alone," and the first recorded ascent by Captain Samuel Willard and fourteen rangers in 1725 made it a useful lookout point on their patrol for Indians. In the nineteenth century, Mount Monadnock was beloved of Ralph Waldo Emerson, who called it the "constant giver," and of his friend, the poet-philosopher-naturalist Henry David Thoreau. "On the tops of mountains," Thoreau wrote in his book *Walden*, "as everywhere to hopeful souls, it is always morning."

Monadnock also has a special place in the lives of Lynn Riddiford and Jim Truman, who set out to climb the mountain on a Sunday morning in 1969. The autumn was fighting back winter, when the "roots and rocks" trails of New Hampshire would be more treacherous due to mud, snow, and ice—and for now, fall was winning handily. As they drove toward Monadnock, the views were spectacular: The deciduous forest mixed bright-yellow foliage with rich reds and vivid oranges. The air was crisp. The sky was blue. An hour and a half from Cambridge, Lynn, thirty-three, and Jim, twenty-four, had nothing to worry about. Up there, on the mountain, there was no chance they'd meet anyone they knew.

Or so they hoped. Just three years earlier, Lynn had become the first-ever female tenure-track professor of biology at Harvard. There were six female professors in total at the university then—only one tenured in the entire Faculty of Arts and Sciences—and they were expected to be on their best behavior. Jim's arrival in early September 1967 jeopardized all that. By wintertime, the supervisor and her doctoral student were falling in love. It was understood that male mentors often had romantic relations with their underlings, understood and tolerated, but this wasn't supposed to happen the other way around. For a year and a half now, Jim and Lynn had been going steady, but no one at the university knew their secret. No one, except Carroll Williams, the department chair.

Fastening their backpacks, they held hands and began walking up the side of the mountain. Lynn Riddiford (née Moorhead) had grown up on a dairy farm in Illinois, surrounded by animals. Her father was the dairy farmer son of a dairy farmer, and her mother the English teacher was the daughter of a dentist. By the time she was three, she would "help" her father milk the cows, every third Sunday night, and when she got tired he'd set her on the back of an old horse named Methuselah and send her home across the field; the horse would walk into the barn and little Lynn would call for her mother until she came to help her down. Lynn and her sister were brought up on solid Christian values of hard work, faith, and modesty. Though they had to take home economics in grade school, rather than agriculture (for farm boys) or shop (for the boys who lived in town), life was a happy blend of wholesome and exciting, raising Holstein dairy calves, riding horses, and participating in girl's 4-H cooking and sewing.

Jim, on the other hand, grew up in a middle-class residential neighborhood on the east side of Akron, Ohio. His dad was a mechanical engineer at Goodyear Aircraft, and his mother stayed at home and took care of the kids. There wasn't much to read in their house, beyond the *Encyclopedia Britannica* and the Bible, but Jim was constantly borrowing books from the public library a mile and a half down the road. There was a house trailer his father had built that figured prominently in the family's summers, and an older brother and a younger sister. There was regular attendance at Mass every Sunday, meatless meals on Fridays, and four years of Catholic high school, run by the brothers of the Congregation of the Holy Cross.

And for each of them, nine years and a universe apart, there was science. In Lynn's case it began with an article in *Seventeen* magazine. The author was a student who'd spent a summer at the Roscoe B. Jackson Memorial Laboratory in Bar Harbor, Maine. Lynn applied to the same program her junior year, and was one of fifteen accepted nationally to attend for ten weeks that summer. Fifteen-hundred miles from the farm, it was the farthest away she'd ever traveled, and when she saw New York City en route she was awestruck. She did so well, she was asked to return the following summer. The biologists she'd met at Bar Harbor made science come alive: Dr. Leroy Stevens supervised a project involving stem cells, and Mr. Frederick Avis held her hand as she observed the effects of hypoxia on the prenatal development of mice. Lynn was electrified. In her senior year, she was disappointed to be chosen as salutatorian (the principal's son was valedictorian). Even so, she had set her eyes on East Coast colleges; she wasn't going to settle for the local Rockford College where her mother had gone. By the time Radcliffe College offered her a full scholarship, she was ready to spread her wings.

Jim had fallen in love with science thanks to Roy Chapman Andrews, the pioneering paleontologist and adventurer who wrote thrilling accounts of his discoveries of dinosaurs for young readers. In the city park two blocks from home, and during summers on the shores of Lake Erie, Jim became a collector of stones, shells, butterflies, beetles, Lincoln head pennies. Once he made a brain collection for a grade-school science fair, dissecting the brains of frogs, lampreys, and hamsters. Rather than simply eat the rabbits and pheasants his father shot, Jim took a mail-order course from the Omaha School of Taxidermy to learn how to tan skins and mount them. (He was twelve.) Bumping into a parasitic wasp with a three-inch-long abdomen in the trailer park at sixteen, he seized on Frank Lutz's classic *A Field Guide to Insects* and was hooked. He enrolled at Notre Dame as a National Merit scholar with a concentration in biology.

Lynn's path wasn't so straightforward. In her junior year at Radcliffe, she had applied to vet school at Cornell and was called in for an interview. Entering a room with six men sitting around a table, she stood before them to answer their questions, since there was no chair. "Can you cook and can you sew?" she was asked. "Yes and yes," she answered. "And why do you want to know?" Their vet students always marry their fellow female vet students, they explained to her, so they had to make sure they were well taken care of. When Lynn said she wanted to work on cows and horses, they said she was too small and would do better working on cats and dogs. In the end they admitted her, but she politely turned down the offer. She already knew that doing biology was going to be more exciting.

After completing a doctorate at Cornell, getting married and divorced to a man named Riddiford, and spending a few years teaching biology at Wellesley, Lynn was ready to return to Harvard.* In fact, almost a decade earlier, someone there had captured her imagination during a speech to her undergraduate Radcliffe pre-med and science club. The year was 1957, and Carroll Williams had spoken to the young women about his own early fascination with bugs, growing up on the banks of the James River in southern Virginia. Specifically, he had noticed that the silk moth pupa lay dormant during the fall and winter. Joining Harvard as a professor years later, he chose to study this condition of suspended development, which scientists called *diapause*. That's when he made a serendipitous discovery: The abdomen of the male silk moth contained a substance that prolonged

---

* Before Wellesley, Lynn had followed Riddiford to Harvard, where she did a post-doc with the protein scientist John Edsall, while Riddiford pursued a doctorate in applied physics.

its life by staving off metamorphosis! A British scientist had already named the substance "juvenile hormone." Now, in front of a room filled with wide-eyed students—and a gimlet-eyed Lynn—Williams spoke glowingly of what he called the "golden oil." No one knew what it was made of, said America's greatest insect man, but it was the magic molecule that kept creatures young.

Hearing such dramatic words, Lynn raised her hand to ask a question: Would this same "golden oil" stop metamorphosis in a tadpole? Williams paused a moment, said he didn't know, and invited Lynn to join his lab to find out. Years later she'd joke about killing untold tadpoles by injecting them with the "golden oil" from the silkworm; alas, it did nothing at all to keep them young. Still, Lynn was enchanted. Her senior thesis the following year used Williams's procedure to extract "golden oil" from day-old baby rats and bovine organs and test it on Polyphemus, a giant wild silk moth that looks like a spider with wings. This time, there was an effect—it stopped metamorphosis entirely! Whatever juvenile hormone was, it existed in the mammalian thymus, and adrenal cortex, and as another student discovered, even in the human placenta. Williams wrote it up, and the results of Lynn's first experiment were published in *Nature*.

And so when an offer came to join Carroll's lab as a research fellow, Lynn was thrilled. Soon she was appointed the first-ever woman on the biology faculty. It was the fall of 1966, and Lynn was kindly asked to take her meals at the Ladies Dining Room—reserved for the wives of Harvard professors—rather than the main dining hall at the Faculty Club. Nor, as a woman, was she allowed into the undergraduate library. But the department itself was friendly (apart from the fire ants from E. O. Wilson's lab beneath hers who kept climbing up and stinging her caterpillars), and the new molecular techniques were amazing. Over in Notre Dame at the very same time, Jim had been adopted as the mascot of a mosquito genetics lab. It's head, a man named George Craig Jr., had opened up his mind. When he graduated, Jim decided that the most exciting way to continue opening his mind was to come to Harvard, to sit at the foot of the great Carroll Williams.

Up on Mount Monadnock on that Sunday morning in 1969, Jim and Lynn might have laughed at the shock Jim received when he heard he'd been admitted to the doctoral program to work, not with Carrol Williams, but with the young assistant professor Lynn Riddiford who'd interviewed him the penultimate spring. They might have laughed, too, about how quickly he began to woo the dark-haired, blue-eyed beauty who shared his passion for moths. They might have laughed at all this, had they not rounded a bend on the trail and bumped directly into half a dozen Harvard students. One of them, alas, was an undergraduate working in Lynn's lab.

After a year and a half of discretion and subterfuge, their secret was, as Jim later remembered, "toast."

2.

RUMI

I met Yaeli by chance. There was no secret involved, no illicit love affair, or sneaking around. I was nearly forty, a perennial bachelor, and even my folks were beginning to lose hope. One day my mother sat me down to explain to me that waiting for Miss Perfect was a fool's errand since the strings of the heart change their tension all the time. It was right about then that a childhood friend called me out of the blue one evening: "There's a girl I swam with in the ocean today that I think you should meet," he said. Soon I learned just how little basis there was for his recommendation: "I don't know anything about her besides her name, but she has a beautiful butterfly stroke." Sensing some skepticism, my friend added, "Don't you know? Swimming tells you all you need to know about a person's heart!"

It was a strange endorsement, but perhaps as good as any other. And to be perfectly honest, I was losing hope in myself, too. I'd failed so miserably at love that I was open at this point to just about anything. That same week I'd moved to a new apartment, and it turned out the butterfly girl—Yaeli—was my neighbor. Little did I know on our second date, stumbling clumsily after her over rocks and sea squirts into the Mediterranean, that I was about to embark on the swim of my life.

Now, five and a half years later, we already had two children and were expecting our third. The closer the day came, the more I turned philosophical. I'd long graduated from butterflies and Catherine the Great. Now I was obsessed with a more abstract concept—identity. And I couldn't stop listening to music inspired by a mystic from the thirteenth century.

When the Mongols invaded Central Asia in those medieval times, a jurist and theologian from Balkh in latter-day Afghanistan named Bahā ud-Dīn Walad took his family and a group of disciples and headed west. According to an account not entirely agreed upon by scholars, his son, Jalāl ud-Dīn Muhammad Balkhī, who was just a young boy, was walking a few steps behind him when in the city of Nishapur they encountered a famous Persian mystic poet. Immediately recognizing the boy's spiritual preeminence, the poet said: "Here comes a sea followed by an ocean."

Later, Bahā ud-Dīn Walad settled the family in the westernmost territories of the Seljuk Sultanate of Rûm, and when he died, Jalāl ud-Dīn Muhammad Balkhī inherited his father's position as molvi. Soon the son became a noted juror and teacher, issuing fatwas and giving sermons in the mosques. The province of Anatolia had once been ruled by Byzantine, the Eastern Roman Empire, hence its name, Rûm, Arabic for "Roman." This became Jalāl ud-Dīn Muhammad Balkhī's epithet, the "Mullah of Rûm," or simply "Rumi."

One day Rumi met the Sufi mystic Shams e-Tabrizi, who had long traveled the Middle East in search of a partner for spiritual enlightenment. According to legend, a voice now said to him, "What will you give in return?" and he replied, "My head!" to which the voice answered, "The one you seek is Jalāl ud-Dīn of Konya." Following their encounter, the two men spent months together without any human needs, transported in deep conversation. They were so enmeshed in one another, it was said, that when they fell at each other's feet no one knew who was the lover and who the beloved. Some years passed, and as they were talking, Shams was called to the back door and was never seen again. Rumors spread that he had been murdered, perhaps by Rumi's own jealous son. He had given his head, people whispered, for the privilege of a mystical friendship as the voice had foretold. Grief-stricken, Rumi began to listen to music and to sing, whirling around himself in abandon. He wrote poems about love like this one:

Through Love all that is bitter will be sweet
Through Love all that is copper will be gold
Through Love all dregs will turn to purest wine
Through Love all pain will turn to medicine
Through Love the dead will all become alive
Through Love the king will turn into a slave!

From a juror and teacher, Rumi had graduated into a Sufi mystic, and in time would become the favored poet of Tajiks and Turks, Persians and Pashtuns, Greeks, Bengalis, and many moderns. I had become one of them.

Rumi had gotten me thinking about identity, about what constitutes a self. From Maria Sibylla Merian and her times, I had learned how intimately selfhood had been tied to the question of where we come from. From Ernst Haeckel and his day, I had seen how thinking about selfhood related to hopes about where we're going in years to come. Both Merian and Haeckel, in their own ways, had addressed puzzles about origins and progress that originated with Aristotle. But so much exciting science had happened since their times, and so much new culture had amassed. I was ready to learn what we know today about metamorphosis. With the help of my friend Jens, it dawned on me that Jim Truman and Lynn Riddiford were probably the two people who knew more about metamorphosis than any other person who had ever lived. For over sixty years, they'd devoted their lives to the subject, interrogating the transformation of life-forms from almost every possible angle. It was Lynn and Jim who would figure out precisely which genes and hormones are responsible for the life cycle, and how they work together to turn a caterpillar into a butterfly. It was they who would first uncover how the brain of a maggot becomes the brain of a fly. Jim and Lynn would also offer a tantalizing theory going back to ancient times about where metamorphosis had come from. So I got on a plane to meet them in their home on a tiny island. Their story, I'd find out, was a love story: a love for living creatures and a love for science. What they discovered really does turn the king into a slave, us self-important humans into humble observers of a larger drama—or at least it should. Without knowing it, Lynn and Jim were living examples of how modern science may or may not be used to answer our most intimate human question: the question of what is a self.

But before we join them again, before we delve into the molecular world they uncovered, and the evolutionary origins they imagined, and the transformations of brains they tracked—before all that, we need to return to the oceans. There, underneath the waves, is where the larger drama begins.

3.

SEA SQUIRT, STARFISH

There are people in this world who think that sea squirts are boring. Just tubes fastened to a rock or a coral, maybe a mangrove root or perhaps a ship's hull. A sea squirt looks like a barrel, with one large opening at its tip and another to the side, down a bit on its trunk; water comes in through one siphon, filtering in tiny planktonic snacks, then goes out the other. There's a heart just below a gut, an anus, a simple intestine. But there aren't any sense organs, no limbs, not even a head. Three thousand species strong, sea squirts belong to a class known as ascidians, the Greek diminutive of *askos*, or "wineskins." In the wild, they just seem to sway through life, with very little going on. People say they're so boring that they don't even have natural predators.

But I believe that sea squirts are wonderful. Their tunics can look like grapes or peaches or barrels or bottles. They come in luscious red or bright orange, technicolor, fluorescent, or translucent. Some are hard as a butterfly's carapace and some even softer to the touch than a slug. And since the cellulose in the tunic can be converted into ethanol, sea squirts might prove a surprising source of biofuel one day. Catching a ride on the hull of ships, or in ballast water or on the shells of mollusks, far-flung species have colonized new pastures from the Mediterranean to the Pacific Northwest. Sea squirts seem completely harmless. Nothing about them gives away just how wild their lives really are.

When a sea squirt egg begins to grow, it first turns into a baby larva that looks a lot like a tadpole. It has an eye and a mouth, and a tapering tail with which it swims. It's also built around a spine: Precisely 215 cells build a structure called a notochord, one that scientists study today to figure out the origin of vertebrate brains. Within hours or days, depending on the species, the larva ends the charade of apparent independence. Swimming away from the light, it heads downward, where it latches onto a coral or rock. In a matter of minutes the tail will be consumed from within, melting away like an ice pop. The notochord is destroyed, its contents devoured by special digesting cells in the adult. Soon the brain is also ravaged. All the cells that can be reused will be retooled and recycled, to serve the growth and form of the adult. Ready to begin its tethered, plankton-filtering life, the headless, eyeless, limbless sea squirt now emerges, victorious. Only barely noticeable teeth within its syphon give any hint that it ever was a completely different creature. The larva has "sacrificed" itself for its grown-up self, disappearing as if in an act of sorcery. Harboring a dark secret, silently, the sea squirt sways gently in the current.

The situation begs a psychoanalyst, as much as an embryologist: In human terms, it would be like a baby eating itself to become an adult. Except that even that impossible scenario doesn't do justice to the weirdness of the ascidians. This baby is absolutely nothing like the adult: It looks like a tadpole, with a tapering tail and a head and a face. Most astonishingly, the adult sea squirt is an invertebrate, but the sea squirt larva is a chordate, like us humans.

People think sea squirts are boring, but their life cycle is otherworldly. Not that fate is always entirely determined; as in axolotls,

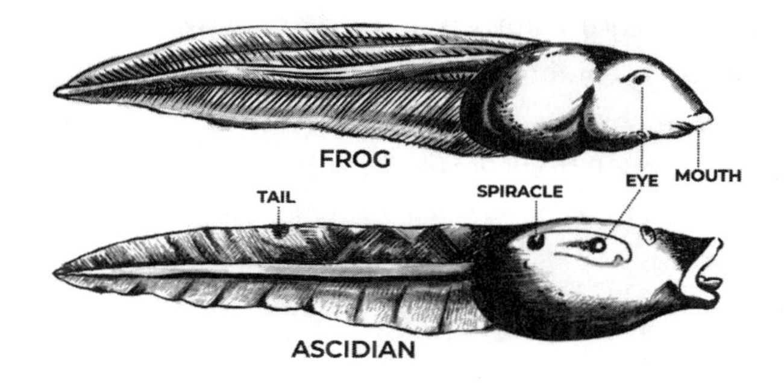

A frog larva compared to a sea squirt larva. Credit: Inna Gertsberg.

there are species of ascidians who forego their final metamorphosis while remaining free-swimming, sexually reproducing larvae their whole lives.* It's as if evolution has let them off the hook, given them a pass. Still, I can't help but wonder: As its tail undulates downward, heading away from the light, is the sea squirt tadpole an independent entity? When its tail begins to shrivel up and its proto-spine to crack, what's going on in the brain it's about to lose: Could it be contemplating the meaning of its sacrifice? The adult sea squirt and its tadpole larva have the exact same genome in each of their cells, but they're neither brothers nor sisters, nor parent and child. What *are* they then: A saint? A schizophrenic? Attempting to answer the question challenges our ideas of agency and selfhood. As much as we'd like to feel the sea squirt from within, we just don't have the language to translate its experience into our own lives.

Leaving aside the mystery of identity, we are also faced with an evolutionary question: Why should ascidians have participated in what looks like a dramatic reversal? Why start life out as a chordate to become an invertebrate in the shape of a sack? The turn-of-the-twentieth-century taxonomist and fish specialist David Starr Jordan thought he had the perfect answer: The sea squirt had once been a "higher" fish, which, due to its own vices, had unfortunately

* Members of this class are called *larvaceans*, and are transparent planktonic creatures less than half an inch long.

been "degraded." The culprit was laziness, or what Jordan and his cowriters called a combination of "dependence," "inactivity," and "idleness," one imagines with a sneer. Alas, "bad habits" are no longer fashionable as explanations for the fortunes of wild animals. So what is going on?

It's a lazy Saturday, and we're folding laundry and doing our best to smile. The last few weeks have been tough: Yaeli aching, the kids complaining, and I—nearing my second half century—more engrossed with my Rumi music and myself than I care to admit. A shrink friend of mine tells me that at my age many men go through a kind of disassembling, that their ideal self is slowly replaced with a more accurate self-understanding, and that seeing yourself more clearly isn't always that much fun.

Later that day I learn that it was a Russian named Alexander Kovalevsky who first connected between the free-swimming tadpole ascidian and the sessile sea squirt, affixed to coral rock. This was in 1866, and both Ernst Haeckel and Charles Darwin were astounded. In evolution ascidians were ancient, going back 540 million years, to the Cambrian. "Thus, if we may rely on embryology, ever the safest guide in classification," Darwin wrote in his *Descent of Man*, "it seems that we have at last gained a clue to whence the Vertebrata have derived."

The next day, at the beach, I come across a starfish. Maybe it's their gently tapering arms that make them lovable to me, or maybe it's their proud silence that sways my heart. Maybe I sense in them a kindred spirit, the five appendages splayed in every direction, as if hoping to give the world a hug. My daughter Shaizee loves them too. But I'm pretty sure she'd be shocked to learn the truth about the one we just encountered. Not because this velvety bottom-dweller of the northern seas and Mediterranean is a ruthless predator. Not because it eats members of its own species in the dead of night. What would truly flip Shaizee out about *Luidia sarsii* is a mind-bending fact about this and other starfish: It can remain a child while simultaneously becoming an adult.

Like any other creature, the starfish embryo begins to develop, one cell dividing into two, then four, then eight, then sixteen. At thirty-two they turn into a little ball, called a *blastocoel*. In conventional development, the blastula stage is followed by a gastrula, a ball with cells three layers deep. Each layer is destined to give rise to different parts of the developing animal: The heart and muscles will develop from the mesoderm, the skin and nerves from the ectoderm, the endoderm will eventually produce the lining of the gut. Now cascading molecular signals will direct each cell in each layer to migrate in the embryo until everything is in place, and—*presto*—a creature will be born.

Except that in the starfish it doesn't work like that. When the embryo reaches the stage of the gastrula, instead of one plan continuing to unfurl on the way to becoming one animal, two plans start to unfold side by side. The "fate maps" in one of these embryos begin to sculpt an incipient larva, while the "fate maps" in the other go about producing a young adult. The description may sound simple enough, but consider what just happened: A fertilized egg that somehow contains not one, but two, developmental programs, has kicked both into gear at the very same time. And while the larva initially grows at a faster pace than the adult, the very same creature goes about building two brains, two hearts, and two anuses.

In the meantime the larvae, called *bipinnarians*, grow to be an inch and a half long. On one end of their body are two angelic flappy wings, fleshy and transparent; these help them dive down into streams during the day, and glide to eat algae on the surface at night. The other end of the body is a tangled mess of arms that look like the roots of an upturned vegetable. It may not seem that way at first, but as its name suggests, the larva is now a bilaterian: With a front and a back side, a head and tail and left and right, its body plan is symmetrical, just like ours.

Except all the while, inside itself, this tiny larva has been growing its adult. From a cluster of special stem cells in its inner cavity, an alien has been sprouting within. Slowly, a star emerges, with its own brain and heart, appearing on the tangled roots—salmon pink, or bright orange, a tiny splash of color on the pellucid stalk—like a jewel. Rather than bilateral symmetry, it has radial symmetry—the symmetry of a starfish. Gradually, the five arms elongate, with a fringe of white spines. And in a moment of poetic grace—a truly immaculate conception—the adult detaches from the roots and falls, sinking in the water column, like a rock.

This is what's incredible: Unlike the sea squirt whose larva "sacrifices" itself for the adult, here, the two part ways, each to lead its separate life. The colorful, sexual starfish settles on the sea floor where it will grow a hundredfold, as the angelic inch-and-a-half-long translucent larva just resumes its virginal pelagic life. We're accustomed to thinking that adulthood always follows youth, and that youth precedes adulthood. But starfish offer a radical alternative. Incredibly, in a single organism, the two can persist, side by side.

Our human notions of identity have once again met their limit. The larva gave birth to the adult, but having sprung from the very same fertilized egg, it is not its parent. It is not its brother, either, or a cousin. The genes in each and every one of the cells of the two separate creatures are identical, but two separate developmental programs have produced a relation that we can't quite wrap our heads around. Looking at the two creatures, it's easy to see that they show no interest in one another. If we encountered both together, we'd have no idea they had the same origin. But they are, in a very real sense, the self-same creature. If I didn't know better, I'd turn to Rumi for answers.

"Hungry, you're a dog," the thirteenth-century mystic wrote, "angry and bad-natured. Having eaten your fill, you become a carcass; you lie down like a wall, senseless. At one time a dog, at another time a carcass, how will you run with lions, or follow the saints?" Rumi's answer to his own question was that one needed to ween oneself of the ego. To do that, rather than resisting change, it was necessary to embrace it. Even small change could lead to great diversity of spirit, and therefore great resplendence. Even change that was based on persistence. Change, and the change of change, and its persistence, were the lifeblood of the soul.

There was a reason why my ear had begged for music inspired by Rumi. His poetry was the mystical complement to a new modern science. Gradually, I began to perceive, like a dog rising from its slumber, that it was this science that provided the key to understanding metamorphosis.

4.

EVO-DEVO

Evolutionary developmental biology, or "evo-devo" for short, is a field in biology that compares the development of different creatures to infer how development evolved. Another way to express this is that it's concerned with how changes in embryonic development during single generations relate to the evolutionary changes that occur between generations. Under this name it's a new field, just a generation or two old, but it's preoccupation is ancient. We won't go all the way back, but instead start shortly after Darwin, with a student of Haeckel named Hans Driesch.

Driesch is remembered as the first non-Jewish professor to be removed from his academic post by the Nazis; he was an outspoken pacifist and came to the defense of Jewish colleagues in the face of nationalist discrimination. He's also famous for moving from embryology to natural theology. Known in his day as a brilliant scientist, he became a philosopher, and in later years devoted his energies to studying what he called "the science of the super-normal." Driesch was a genius who, according to some, had lost his marbles. And what led him down this path was a tiny sea urchin egg.

Back in 1891, Driesch discovered that when he placed a fertilized sea urchin egg in a vial of water and shook it vigorously right after its first cleavage, the two individual cells, called blastomeres, separated from one another, and each continued to grow. The same was true for the four and then eight cells separated by shaking in the next cleavages. Amazingly, in many instances, each of the dislodged cells developed into perfectly healthy larvae, though at a slower pace than normal and usually reaching only half the normal size.

The reason Driesch was spending his time shaking cells apart was to disprove the theory of a man from his alma mater. Wilhelm Roux, the son of a fencing master, had taken a frog's egg and, when it divided, killed one of the blastomeres by pricking it with a heated needle. The surviving blastomere, he found, developed into half an embryo. Roux was sure this could mean only one thing: that the nucleus of the cell contained only half the determinants of an embryo. From the very beginning, there had to be factors of future development laid out in fixed positions within the fertilized egg. How it worked would go something like this: After one division, factors determining the left and right side of the embryo would divide into two separate cells; with the next cleavage, factors determining the head and tail would be divided up into the four top and bottom blastomeres, followed by factors determining the heart, brain, and limbs in the next cleavages, followed by factors determining specific heart chambers, brain structures, toes, and so on after them. Roux called this his "mosaic theory of development," and ruled out any influence of the outside environment. By an intricate, unfolding tapestry, creatures were determined from within.

But Driesch showed that Roux had been sloppy: He'd failed to separate the remnants of the blastomeres he destroyed with his heated needle, which, it turned out, inhibited the regeneration of all the parts in the adjacent surviving cell. Had he simply separated the cells entirely, as Driesch himself did by shaking, he'd have gotten not half-embryos but entire, half-sized ones. Driesch concluded that the "mosaic theory" was mistaken. To describe what was really going on, he cribbed an ancient word from Aristotle: *entelechy*. What it meant was that each cell somehow had the ability to become an entire organism. How this internal "life force" worked could not be accounted for by the cutting-edge developmental mechanics (*Entwickelungsmechanik*) that both he and Roux had pioneered. And it drove Driesch, a die-hard rationalist, to vitalism, the "super-normal," and God.

Roux and Driesch's rivalry goes back to Ernst Haeckel. Both had been his students at Jena, and it was Haeckel who performed the experiments on which theirs were based. Rather than sea urchin or frog eggs, Haeckel had chosen the eggs of siphonophores, the delicate order of colonial hydrozoa he'd found on the Canary island of Lanzarote, and considered as beautiful as bouquets of flowers. Using a fine needle and a microscope, he divided two-day-old embryos of siphonophores into either two, three, or four groups of cells, showing that each group eventually developed into complete siphonophores. Yes, they were smaller than normal, and it usually took them more time, but nonetheless the magic occurred.

Haeckel performed his experiments thirty years before Driesch, and today they are largely forgotten. The reason for this is partly because Haeckel did not draw Driesch's dramatic conclusion that every cell in an embryo was *totipotent*, meaning that it harbored within it the ability to develop into a full-blown organism. Instead, Haeckel focused on what development could teach us about evolution: Playing around with environmental conditions like water temperature, abundance of light, salinity, and movement, he found that some of the manipulated cells, incredibly, would take on the forms of related species rather than their own. Using the biogenetic law, he sketched out evolutionary trees detailing the relationships. All siphonophores descended from a simple medusa that budded off a hydroid colony, he thought, a theory embraced today for the entire phylum of cnidarians. Haeckel was delighted. Siphonophores, he wrote to Darwin, due to their many forms and shapes, provided "the most excellent demonstration of descent theory."

What Haeckel had done was to think of development and evolution together: How a life-form grows from a single cell to an entire organism could harbor secrets about where it had come from in the deep past. But as embryology became more professional, it tore itself away from the more speculative theory of descent; Haeckel's "apostate students" were more interested in the mechanical How than the evolutionary Why. Roux ended up rejecting his teacher's biogenetic law; and as we've seen, Driesch (whom Haeckel thought might benefit from time in a sanitarium) turned to vitalism, and became a kind of mystic.

In the following decades, genetics and evolution by natural selection became wedded. This neo-Darwinian evolutionary synthesis, as it was called, did a lot to boost biology, placing it on a par with the "harder" sciences of chemistry and physics. But the price that was paid was that development and evolution were completely torn asunder. Treating individuals like gas molecules in a flask, the heroes of the synthesis were not embryologists, but pen-and-pencil geneticists.* Calculating gene frequencies and selection pressures, rather than peering carefully down a microscope at the ectoderm or mesoderm, they paid no attention to the development of individual organisms. To them evolution was not a private affair, but a population phenomenon. And while their equations showed that it could all work—evolution progressing gradually based on the selection of random mutations—they ventured no answers to the question of how a particular creature actually gained its form.

The neo-Darwinians had made a separation where their hero had perceived a unity. Darwin would spend his life debating the relative importance of variation and selection: If there was a whole lot of random variation, selection would need to be like an all-powerful sculptor, giving nature its form; but if variation was not entirely random, if, for example, variation was sometimes a response to the organism's needs in a particular environment, then selection wouldn't need to do quite as much of the work. Darwin swayed in later years toward this second idea, but one thing was clear to him: No evolution could be possible without there first being changes from one individual to

* The three most influential architects of the synthesis were Ronald Fisher (1890–1962), J. B. S. Haldane (1892–1964), and Sewall Wright (1889–1988).

the next. It made no sense to him—and clearly not to Haeckel—to separate ontogeny from phylogeny. Development and evolution were two sides of the same coin.

Darwin's failure to explain *how* change came about in individuals was the single greatest weakness in his theory. Clearly, it had to do with inheritance, but as he wrote in the *Origin of Species*, "The laws governing inheritance are quite unknown." And clearly it had to do with variation, but "variability is governed by many unknown laws." Darwin's ignorance about the workings of heredity and the origins of variation was no coincidence; the two are intimately linked. Never one to give up easily, he offered up some guesses: that variation came about through direct action of the environment on individuals, or through the inheritance of acquired traits. For this last idea, he invented what he called his "provisional hypothesis of pangenesis": Every tissue in the body is filled with tiny inheritance particles called "gemmules," which flow to the gonads during sex. When the body parts in which they reside are used, gemmules are augmented; when body parts are not used, gemmules are depleted. According to this scheme, a body builder would make brawny children, because he'd have many muscle gemmules to hand down to the next generation, whereas the progeny of a monkey who left the trees would gradually lose their tail. But Darwin's hypothesis was just that, a conjecture. No evidence of gemmules was ever found.

Around the time Darwin was writing his *Origin of Species*, a Moravian monk named Gregor Mendel began crossing pea plants in a little garden plot adjacent to the Augustinian monastery in Brno. What Mendel found in these experiments was that plants inherited characteristics from their parents in a curiously mathematical fashion. To explain why this happened—why peas came out round or wrinkled, green or yellow, tall or short in a predictable way—it made sense to speak of "elements"* conducting the business of inheritance. Mendel believed that these "elements" were either dominant or recessive, that each trait was determined by two versions of the trait

---

* In German, *Anlagen*.

(alleles) that were inherited independently from each parent, and that each trait was passed down to the next generation separately from other traits. Were they made of matter? Were they perhaps a force? Mendel had no clue.

In years to come, Mendel's elements became known as "genes," and brought about a revolution. After chromosomes were discovered in the nuclei of living cells in 1882 (so named from the Greek for "colored bodies" since they could be made apparent to the eye by special dyes), it was another of Haeckel's students, Oscar Hertwig, who figured out that certain cells in sexually reproducing creatures had to do a special kind of cell division. These cells didn't just replicate themselves, the way for example a skin cell replicates, in a process called *mitosis*. Instead, they also engaged in reduction, a process he called *meiosis*. In each of the sexes, meiosis would halve the chromosomes of certain cells in order to produce egg and sperm that could then unite. Where we come from and where are we going were finally being unlocked: With the correct number of chromosomes carrying a new combination of genes, inherited from mom and pop, a new creature would come into the world.

Meanwhile scientists began to believe that genes were very real things. No one had actually seen them, but they resided in the nucleus of cells in particular places on particular chromosomes, and the behavior of chromosomes during meiosis explained why genes were inherited according to Mendel's laws. This was called the Sutton-Boveri hypothesis, named after the American Walter Sutton and the German Theodor Boveri, who published their findings independently in 1902 and 1903. Strangely enough, in 1905, four years before Mendel's "elements" were renamed "genes," the science of heredity was given a name—*genetics*. And what geneticists increasingly showed was that inheritance was a game of chance. Like cards dealt in a poker game, genes were randomly shuffled during meiosis and sex, and they also mutated. Sometimes the shuffling and mutations were helpful, and sometimes they were deleterious. Turned into mathematical abstractions, their ebb and flow in populations could trace the evolution of life.

Genetics and evolution had now become respectable sciences, but the holy grail since Aristotle—How do individuals develop?—remained a mystery. By what alchemy, from a tiny egg, do they grow countless cell types, which migrate as if possessing a compass to build organs in just the right place? If it was inherited change that was important, then it made sense to look directly at the gametes. Back in the nineteenth century, the same August Weismann who had determined that axolotls could be coaxed into adulthood had asked a simple question: Where in the developing embryo are the cells that will turn into sperm and egg? What he found, looking at a medusa, was that the gametes came from cells that were physically separated from the cells that made up the rest of the body. This was the strongest evidence yet that changes acquired to the body couldn't be inherited.

But the very man who coined the term *genetics*, the Englishman William Bateson, had already noticed something strange: When things went wrong in development, they often went wrong in similar ways. That's why human babies born with a sixth toe or finger were relatively commonplace (polydactyly, as it is called, occurs in one in every five hundred to one thousand births), as are insects with a double set of wings or legs in place of antennae, turtles with two heads, even (somewhat more rarely) calves with a cyclops eye in the middle of their heads. If mutations were random, why would genetic glitches be anything less than random? Why does faulty development result in the very same mistakes time and again?

Bateson didn't have an answer, but the German embryologist Hans Spemann began to unravel the mystery in the teens and twenties. While recuperating from tuberculosis in a sanatorium at the turn of the century, the young biology graduate had read a book by August Weismann. *The Germ Plasm: A Theory of Heredity* preached that with every cell division, genes are chopped in half: If you start with one cell that has say, heart, liver, brain, and gut genes, the first division might create a cell with liver and brain genes, and another with heart and gut genes, but a further division would produce unique genetic cells that would then go on proliferating to accomplish their mission: the first for the creation of a heart, the second for a brain, the third for

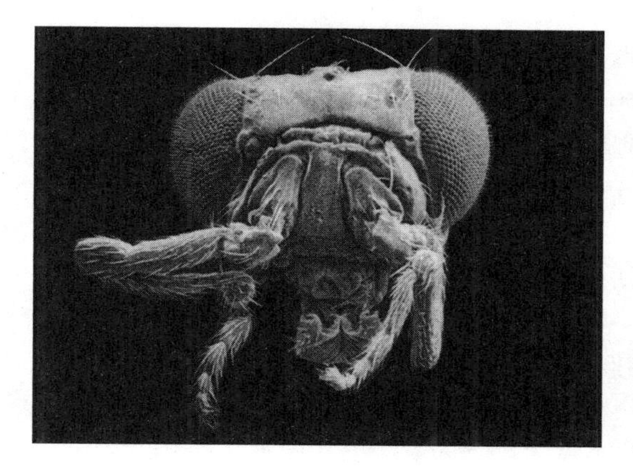

A fly born with legs on its head rather than antennae. Source: F. R. Turner and A. P. Mahowald.

a liver, and the fourth for a gut. Like a search for the self by a process of elimination, each generation of *daughter cells*, as Weismann called them, would have fewer and fewer genes, until a cell could determine what it really was.

This was essentially a new version of the "mosaic theory" propounded by Wilhelm Roux—and undermined by Hans Driesch. But could the differing results of their experiments have been due to the fact that one observed sea urchins while the other looked at frog eggs? Maybe Roux wasn't as bad a scientist as Driesch made him out to be. After all, it seemed ludicrous to believe that you could get a whole organism from tiny bits of embryos. Maybe it only worked for Driesch since sea urchins were so much simpler. No longer suffering from tuberculosis and now a professor, Spemann was gaining a name for himself as a talented microsurgeon, and a master maker of tiny instruments—especially glass rods rendered into silk-thin strands over a Bunsen burner and used as micro-fingers to manipulate cells. Choosing salamander eggs for their unusual size and transparency, easy incubation, and availability, he went to work. Whatever was responsible for the gap between Roux and Driesch's half-embryo and half-sized embryo theories, Spemann hoped to see it with his own eyes.

Except that salamander eggs, as wonderful as they are, soon proved to have problems of their own: When he shook them, à la Driesch, they were sticky and wouldn't come apart. Pacing in his living room with his newborn baby girl on his shoulder, Spemann knew he needed a new instrument, not a fingerlike probe to pick the embryos apart, but a kind of noose to tie it off. What about . . . *a baby's hair*! Quickly, he grabbed a pair of scissors, cut one off Margrette's head, and headed for the lab.

What Spemann went on to accomplish would transform biology. Placing a salamander egg that had just cleaved into two under the microscope, he carefully used tweezers to tie his baby's hair around the budding embryo. Tugging at both ends, he made sure the cells became separated without becoming ruptured. If Weismann and Roux were right, each cell would now develop into something different: one into a tail, perhaps, the other into a head. But what Spemann got, after a few days of biting fingernails, were two perfectly normal salamander embryos. Genes were not being divided up, at least not at this early stage. Clearly, cells were retaining all they needed to create an entire salamander.

Driesch had been close to the truth, after all, but the solution to the mystery of development remained out of reach. While Spemann was able to get perfectly developing salamanders by tying off the embryo at the two-, four-, and eight-cell stages, after the eight-cell stage the ability was lost. Now either the salamander developed improperly or didn't develop at all. Somehow, the cells' fate was becoming determined at a certain point. But who, or what, was responsible?

He pressed on, looking more closely at the structure of the egg. In particular, he noticed that a gray crescent appeared shortly after fertilization, just opposite from where the sperm had entered. Soon Spemann realized that when this crescent encompassed both daughter cells during cleavage, two perfect salamanders resulted. But when cleavage was unequal, and only one daughter cell received the entire crescent, the other daughter cell just developed into a blob of tissue he christened a belly-piece.* "Without My Gray Crescent to Embrace Me I Am Just a Belly-piece," Spemann proposed a song, with characteristic German humor. Something in the gray crescent was essential for development. But what? He didn't know.

---

* *Bauchstück* in German.

Observing sponge embryos more than half a century earlier, Haeckel had laid the groundwork for an answer. After the fertilized egg goes through a series of cleavages to form a spherical mulberry-shaped mass of densely packed cells called a *blastula*, he saw, it begins to fold into itself to create the different layers from which its organs will stem. This *gastrula*, as the folded structure was called, now takes on an ellipsoid or egg-shaped form by more strongly growing in one direction, allowing the central cavity to expand, and breaking through an opening at one pole that will become a mouth or an anus. Haeckel called this process *gastrulation*, and posited that it was ancient—that, in fact, gastrulation united all complex animals* from sponges to humans in one common tree of descent.

Building on this knowledge, Spemann used a clever trick to demonstrate that cells before and after gastrulation were dramatically different. Transplanting cells that were supposed to turn into skin cells from the *early* gastrula of one species of salamander into the area of prospective neural cells in the *early* gastrula of another, he took advantage of the fact that each had slightly different pigments to follow their fate under the microscope. What he found was that the skin cells turned into nerve cells, switching their identity. This was strong evidence that Roux had been wrong to believe that only the determinants within the cell controlled its fate, rather than the relative position of a cell in a developing embryo. But when Spemann took prospective nerve cells from a *late* gastrula of one salamander and implanted them in the prospective skin tissue of a *late* gastrula on the other salamander, they stuck to their guns and went on to create nerve tissues. The same was true in the opposite experiment: Prospective skin cells from one newt implanted in the region of prospective nerve cells in the other just went about their business forming more skin structures as if nothing had changed. Like babies who can become anything in life, cells were totipotent before gastrulation. But after gastrulation, like inmates serving a life sentence, cells were consigned to their fate.

* Coined *metazoans* by Haeckel in 1874.

At around that time, Hilde Proescholdt, the middle daughter of a soap factory owner, joined the lab and fell in love with Spemann's chief assistant. By now Spemann had realized that the gray crescent gives rise to the cells that initiate gastrulation, and that those cells formed a dorsal lip, called a *blastopore*, that was very special. In all vertebrates, from salamanders to humans, it was the blastopore that built a notochord, literally giving a creature its spine. Hilde Mangold (now married to Otto Mangold) therefore decided to take a closer look at the blastopore. There was a dimple there, tiny but recognizable, and she removed and transplanted it—259 separate times—into a host embryo (once again of a different species with a different pigment so that the fates of the different cells could be traced). As she showed in her 1923 doctoral dissertation, the result was nothing short of incredible: The grafted tissue led to the formation of an extra notochord. And around this extra notochord a second salamander embryo now formed, directly facing its host's embryo, like two conjoined twins. The dimple was the only tissue from the early gastrula whose fate had already been determined, but it hadn't performed its magic alone: Mangold showed that it recruited host tissues in its new neighborhood to help it, practically hijacking them to change their cell fates.

Spemann looked on in astonishment. Somehow the tiny dimple knew how to orchestrate the ectoderm and mesoderm layers to form a notochord and the back-front, left-right, top-bottom axis that gives a living creature its form. Somehow it steered the cells migrating through the back part of the dorsal lip to give rise to skeleton and muscle, inducing the overlying cells to become ears and eyes, brains and spinal cord—step-by-step creating the self. For lack of a better word, Spemann named it, somewhat fuzzily, the *organizer*. And he and Mangold published their results in 1924.

Eleven years passed and Spemann was feeling blue as he traveled to Stockholm to receive a Nobel Prize. The farther north the train sped, the more concertedly he willed the light of the snow beyond the window into his crowded thoughts. A scientist's job was to put together pieces of truth, like fragments of a shattered vase. But his vase seemed cracked to him. Months after they published their joint paper, Hilde Mangold had been killed by the explosion of a gasoline stove; she'd been warming up milk for her baby, the one whose hairs, together with those of Spemann's own child, had made many an experimental noose. Although the rigid bylaws of the Nobel Committee precluded anyone from winning the prize posthumously, Spemann always insisted that he had received the Nobel thanks to "The Experiment of Hilde Mangold." While many organizers would be discovered in the years ahead, the most famous of them all remains the "Spemann-Mangold." And Mangold's dissertation remains one of the few ever to have won anyone a Nobel Prize.

Spemann died in 1941, a complicated figure. Before leaving this world he proposed a "fantastical experiment": transferring an adult nucleus into an egg cell, an idea that would one day produce Dolly the sheep, the world's first cloned animal, in the 1990s. But while he touted a top-down model of control and regulation, he continued to believe that a strictly mechanical description of life was not enough. One of the world's finest experimentalists spoke of "nature acting as an artist," and he went to his grave convinced that there is mind in matter and matter in mind. "For hours I could sit and watch an egg cleave," he told an American audience,

> or the cells move through the blastopore, or the neural folds rise and fuse to form a neural tube, or the tailbud embryo move in circles within its membrane by the beat of thousands of cilia. Some scientists see the beauty and order in the structure of an atom, or in the sweep of the Milky Way and the swing of the Pleiades, or in the homeostasis of blood sugar, or in the adaptations of bird for flight, or in the color and patterns of insects; but I found them in the embryo.

It was a moral lesson, beside an aesthetic one. "And what a glorious society we would have," Spemann preached as the Second World War drew nearer, "if men and women would regulate their affairs, as do the millions of cells in the developing embryo."

Spemann and Mangold left behind many unanswered questions. What in the organizer gave it such power? Was it some kind of genetic factor or factors? Most of all, could the organizers themselves evolve? In years to come, these questions would be at the heart of a new science called "evo-devo." In the meantime, across the Channel, an Englishman with a funny name and an obsession with kissing bugs would crack the mystery of metamorphosis wide open.

5.

WIGGLESWORTH

After his death, his children lamented that they hadn't been insects. "The father of modern insect physiology," people called him, and to his boys that seemed just right. The mystery their dad faced was far too great not to conquer, absorbing all his attention, and so he gave it his life. On all accounts punctilious as well as painfully proper, if not the best father, his name was Vincent Brian Wigglesworth, FRS.

It was an unusual life, certainly, but the mystery itself was even more bizarre. How could a lowly caterpillar disappear into a chrysalis and emerge a flying god? Growing up at the turn of the twentieth century in the sleepy hamlet of Kirkham in Lancashire, Wigglesworth discovered the books of Jean-Henri Fabre, a recluse Frenchman from Sérignan, France. Known as "the Homer of Insects," Fabre recounted in *The Life of the Caterpillar* untold stories of insect adventures that first opened young Vincent's eyes.

Carefully observing the life cycle of the giant peacock moth, Fabre had perceived that the slumber the caterpillar seemingly falls into is nothing but a ruse. The cocoon, in fact, is the site of almost unimaginable activity. But rather than simply adding legs and wings and antennae to its silky body, hardening its carapace, shrinking and remodeling, the entombed caterpillar does something more dramatic—it melts. When Fabre pricked the pregnant pupa with the blade of a knife, it was a syrupy goo that emanated, gushing slowly like lava from a volcano. Gone were the eyes, the mouthparts, gone was the head, and all the rest. The caterpillar had literally come undone, disassembled, only to build the imago afresh.

Fabre was a god-fearing man, and to him this seemed a miracle. Later in life, he would write to Darwin that he'd have no truck with evolution, judging its tenets a secular inanity. "I am sorry that you are so strongly opposed to the descent theory," Darwin replied in 1880, "I have found the searching for the history of each structure or instinct an excellent aid to observation."

To the shy Wigglesworth, Fabre's miracle was a mystery to be solved: How does the meltdown occur? How does the creature reassemble? What are the factors controlling molting and metamorphosis?

When he turned eighteen, Wigglesworth enrolled to read biochemistry and physiology at Cambridge (the émigré Vladimir Nabokov would arrive the following year). Taking a medical degree, he moved to the London School of Hygiene and Tropical Diseases and began to spread his wings as a medical entomologist. First to Nigeria, the Gold Coast, and Sierra Leone, armed with a self-contained laboratory complete with a bed and bath (and microscope), and a determination to combat a disease spread by the bite of the tsetse fly. Then it was to India, Java, Burma, Malaysia, and Ceylon, where he was present in 1934 during the lethal malaria epidemic that killed more than one hundred thousand. All the while, Wigglesworth was learning the secrets of the insects, a million-times-a-million-times-a-million in number, four-fifths of the living species on Earth.

It was already clear to him that combatting pests to mitigate disease depended first on understanding the insects themselves. And physiology was where the answers were to be found. So he used his position to conduct truly "useful" as opposed to "futile" research, a distinction he thought more meaningful than the usual "pure" versus "applied." The result was a book that got him elected to the Royal Society. A book that dealt, in part, with the birth of a butterfly.

He thought he knew what the problem was. A butterfly has wings that can only be powered by an elaborate musculature. The muscles, in turn, demand a respiratory system to bring them oxygen. Respiration is useless unless the air-filled tubes called tracheae can deliver oxygen to the muscles. Without hemolymph, the insect equivalent of blood, nutrients can't reach their destination, either. But when the caterpillar melts down in the pupa, neither its muscles nor its breathing organs nor its heart or brain remain as they were. And since a butterfly looks nothing like a caterpillar, and lives in a different environment, new kinds of organs suited to new functions need to be constructed from a gooey mush. After all, the butterfly will eat different foods as an imago, so will need entirely new mouthparts, and a new gut. It will fly and land and take off again, so it will need elongated legs besides a pair of hardy wings. It will mate, so will need a set of compound eyes, antennae, and genitals. To orchestrate its new behaviors a new brain will be necessary. Somehow, through a process of massive destruction and reconstruction, a new nervous system will have to be built.

In *Principles of Insect Physiology* Wigglesworth put his nose to the ground. He knew that many insects, like mayflies and cicadas, undergo only an incomplete metamorphosis. Never becoming a pupa like beetles, moths, or butterflies, nor changing their shape dramatically, they grow from nymph to nymph, until their wings and new eyes and sex organs appear in the final molt. Molting allows for growth in size, metamorphosis for change of form; clearly the two were not *synonymous*. But Wigglesworth began to think of them as *homologous*. After all, butterflies and moths may seem a world apart from mayflies or cicadas, but, however they went about developing, evolution had already placed in both the recipe for becoming true adults.

Wigglesworth now saw that his childhood hero Fabre had been wrong: The pupa was not, in fact, a bunch of goo and mush. Instead, everything inside had its place, but it was extremely delicate; treat it with anything less than utter gentleness and it would burst. If you did treat it delicately, you'd see that what were termed "imaginal discs" were cells that had been born early in development from the ectoderm, only didn't decide what they would be when they grew up. Instead, they remained apart, undifferentiated in a kind of sac, until the caterpillar was ready to pupate. That's when they'd wake up to form new eyes and wings and legs and genitals. It was as if the insect had two separate developmental programs that lived side by side but marched to a different clock.

But how could Wigglesworth figure out what was really happening in metamorphosis? In *The Voyage of the Beagle*, Darwin had reported being bitten by a kissing bug near the river Luxan in Argentina: "It is most disgusting to feel soft wingless insects, about an inch long, crawling over one's body," he wrote. "Before sucking they are quite thin, but afterward they become round and bloated with blood, and in this state are easily crushed." One hundred twenty years later, scholars would suggest that the kissing bug (*Rhodnius prolixus*) gave Darwin a horrible disease of the heart and intestines called *Chagas*, which eventually turned him into an invalid.* But it also gave Wigglesworth an idea.

Soon he was turning the blood-sucking, incompletely metamorphosing kissing bug into his own loyal guinea pig. *Rhodnius* could be kept on a shelf like a reagent, and survive without eating for months. It was also known for its fifth quick, and final, molt: the wings appearing suddenly, the larger chest, and compound eyes, the full-blown genitalia. There had to be some signal that brought about such dramatic changes so rapidly, a message arriving to its cells, telling them: "Begin, differentiate, there's no time to lose!" Wigglesworth already

---

* Members of the Reduviidae family are sometimes called "vampire" or "assassin" bugs, and in the Latin Americas are known by the further names *barbeiros*, *vinchucas*, *pitos*, *chipos*, and *chinches*. The kissing bug spreads its pathogen, a wormlike protozoa called *Trypanosoma cruzi*, by defecating near the site of the bite. Unsuspecting humans scratch the area, unwittingly assisting the pathogen to swim into the bloodstream.

knew from the pioneering work of a biologist named Stefan Kopeć in Warsaw that it couldn't be the nervous system that delivered the crucial missive. And so he went searching in the bloodstream for a "hormone."

Hormones had recently become household names, following the discovery of the human gonad-stimulating and other pituitary and thyroid hormones and their role in fashioning breasts and genitalia and muscle mass and voice pitch suddenly at puberty, transforming girls to women and boys to men. Why, Wigglesworth wondered, couldn't the same be true of insects? Day in and day out he'd travel from his home in Beaconsfield to his lab in London, break at a quarter to one for lunch, then return until evening. Everyone from his children to his students would remember Wigglesworth as a cold, reserved man, without emotion. But in reality Wigglesworth's heart was about to burst.

Wigglesworth knew that the kissing bug went through five molts. He knew that the first four of these molts produced slightly bigger nymphs, and that only the fifth turned into a true adult, with wings and compound eyes and genitalia. He knew that the kissing bug drank blood and that its molts were perfectly timed to its meals, beginning at some critical period after the nutrients in the blood had enough time to be digested. Knowing all this, Wigglesworth devised a plan: a careful series of decapitations. Using scissors and thread, he took a kissing bug in its third molt and cut off its head, just after the critical period. Sure enough, it continued to digest its blood meal, to excrete waste, and it went on to molt. But when he decapitated another nymph before the critical period, although it survived for nearly a year, it never molted again.

This was important. To Wigglesworth it signaled a clear and stark conclusion: A hormone released in or near the head was being carried through the hemolymph to the rest of the body, somehow triggering it to molt. When the head was removed before the body got the message, molting failed; but if enough time had passed, even without a head the bug would molt. If this was true, hemolymph from an insect that had passed the critical period should induce molting even in an insect that had been decapitated too soon. And so he beheaded two nymphs, one before and one after the critical period. By joining their decapitated bodies with the help of paraffin wax, he let the juices flow freely between them. The more recent decapitee molted, unsurprisingly, but lo and behold, so did its neck-to-neck mate. Clearly the mixed hemolymphs had delivered the hormone! But where, exactly, was the hormone coming from?

It had to be in the bug's head, Wigglesworth thought, like the pituitary gland in humans, or somewhere close by at least, innervated by nerves coming from the head. Dissecting with great care, he found an organ in the kissing bug's head called the *corpus allatum*. No one understood its function before him, but now Wigglesworth saw that during the fourth molt its cells divided with great flair. Curiously, during the fifth and crucial molt, when the nymph becomes an adult, the activity in the *corpus allatum* ceased entirely.

Another peculiarity arose when Wigglesworth decapitated fourth-stage nymphs, which had passed the critical period, and conjoined them to fifth-stagers that had not: The fifth-stage nymphs did not turn into adults. They just molted into larger nymphs. On the other hand, if he glued a first-stage decapitee to a fifth-stager that was in the process of molting, the headless nymph turned into a tiny precocious adult. If only a molting hormone were involved, it should have turned into a second-stage nymph.

Suddenly Wigglesworth got it: This was a game of activation and inhibition! There must be *two* factors in action: a molting hormone and an inhibiting one. Cutting up butterflies and dragonflies, grasshoppers and locusts, mayflies and moths, helped strengthen his line of reasoning. Gory amalgams Wigglesworth made with two bodies of separate creatures conjoined to a single head of a third proved the point. The *corpus allatum* was inhibiting the change to adult—that's why it shut down at the fifth stage. Wigglesworth christened its product: "juvenile hormone."

The reserved man from Lancashire had invented a field, and become famous. A poem about him by John Updike appeared in *The New Yorker* in 1955, and he'd be knighted in later years. And yet Lynn Riddiford and Jim Truman's department chair, the one with whom they deposited their special secret, would pretty much take it from there. Working at Harvard on the giant silkworm moth, Carroll Williams figured out the details of the system: that bead-like strings of cells around the trachea are what secrete the molting hormone, named *ecdysone*.* That the production of this hormone is controlled by a master hormone—which he called "the brain hormone"—emanating from organs in the brain called *corpora cardiaca*. That the brain hormone is carried in the hemolymph to the prothoracic gland, where it gives the order to start producing ecdysone. That each of the nymphal molts is due to a pulse or more of that hormone, jump-started in special neurosecretory cells in the brain, which send their signal to the *corpora cardiaca*, which sends its signal to the prothoracic. That the neurosecretory cells, cleverly, take their cues from the environment: a stretched stomach full of blood in the satiated kissing bug, for example, or warmer temperature as seasons change in the silkworm moth. That's how the life cycle of insects connects to the world around them.

Wigglesworth had been right about the *corpus allatum*: It was truly the source of the third hormone—his juvenile one—the absence of which allows for the final, dramatic metamorphosis. But it was the American with the Tidewater accent, Williams, who beat him to it, leaving him feeling a bit like a kissing bug without a head.

Like so many discoveries, this one involved a lucky break. A Czech collaborator of Williams's, Karel Sláma, arrived at Harvard with several jars of fire bug larvae. But for some reason, when they were left in their jars, the larvae never turned into adults, the way they had in Sláma's lab in Europe, and the two men were stumped. Then they noticed that the same was true when the larvae were placed on copies of *The New York Times*, *Boston Globe*, and *Scientific American*, and the penny dropped. Unlike European paper mills, American mills

---

* A Japanese researcher had christened them the *prothoracic gland*, and two Germans would later call the steroid hormone *ecdysone*, from the Greek for "shedding," a name that stuck.

used balsam fir to make their pulp, and balsam had developed natural anti-insect defenses that kept the larvae from maturing into an adult. The trees were emitting an analogue of the juvenile hormone.

Soon he and his collaborators had extracted and purified the juvenile hormone, and the media went wild. The "elixir of youth," they called it, and wondered whether, to become Peter Pan, humans should spray it into their nose, massage it into their skin, or simply chug it. In his affable manner, Williams took the hype with a smile. This was all harmful nonsense, good for getting young students excited at a Radcliffe PreMed and Science Society meeting, maybe, but nothing more. And yet juvenile hormone might be used to develop insect-killers much hardier than DDT, even if it was useless for retarding development in humans; after all, insects would find it hard evolving resistance to their own hormones. In time, juvenile hormone–based insecticides would become extremely specific: able to kill just one sort of pest by stunting their cycle, while sparing all the rest.

Jim and Lynn walked down Mount Monadnock sheepishly on that Sunday in the fall of 1969. With their secret discovered, life on campus wouldn't be quite the same anymore. Still, however much Wigglesworth and Williams had pried open metamorphosis, there were many more mysteries to uncover. On Monday, their work would be waiting for them in the lab.

6.

THE MOLECULAR TRINITY

I embark on the 8:55 a.m. from Anacortes to Friday Harbor island. Snowy Mount Baker looks upon the scene from the distance, and the sun is peeping from behind a morning mist. The cold Northwest Pacific waters are silky, almost ominously still. When I walk off the ferry an hour later, two small figures holding hands are waiting for me at the end of the wharf. She's a little stiff-kneed and, wearing large white sneakers, a crimson, three-quarter-length coat, gray wool hat, and bulky dark sunglasses, looks a bit like someone from a witness protection program. He's wearing a green baseball cap and a gray beard, his blue jeans are folded above worn-out walking shoes, and he's missing a few bottom teeth. They smile and shake my hand warmly. At long last, after months of corresponding, we finally meet: Lynn Riddiford and Jim Truman have been married for nearly fifty-five years, and at eighty-eight and seventy-nine, respectively, are known as the world's greatest experts on metamorphosis.

We drive in their small hatchback past giant Douglas firs. A deer jumps into the woods to get out of the way and we stop to watch it; for us they're pests, Jim says, pointing at a double fence he's constructed around a vegetable garden on their three-and-a-half-acre plot. And then, through the trees, barely noticeable until we're right in front of it, a wooden house appears, on a bluff overlooking the waters. The San Juan archipelago lies below us, home of harbor seals and orcas. For millennia before white people arrived, Sooke, Saanich, Songhees, Lummi, Samish, and Semiahmoo American Indians fished these waters in canoes, returning with their catch to dome-shaped willow-reed and grass huts sprinkled on the islands. Inside the Riddiford-Truman home, figurines of walruses and giraffes and owls adorn the living room cabinets, rock and wood and clay mementos from years of international travel. On the walls are photographs of zebras and paintings of alligators; in the kitchen are bowls depicting butterflies, on the chair a pillow sporting horses, on a low, dark-mahogany table neatly stacked *National Geographic* and *Science* magazines. A large, beautifully groomed black cat, Spook, shies away from me, as does Shadow, a mackerel tabby. Light floods through large, wooden, rectangular windows, as if invited to bring with it clarity.

Lynn and Jim's secret became even more public when they announced, immediately after Jim received his doctorate, that they were getting married; unsurprisingly, half of the department already knew. The wedding took place at Lynn's parents' farm in Illinois. That fall of 1970, Jim had become a junior fellow at Harvard, an honor bestowed upon young scholars judged to have extraordinary potential. Already he'd chosen his future path: the biology of insect behavior. Back at Notre Dame, he'd witnessed how powerful biological control could be over behavior: Grinding up the gland that makes sperm in a male mosquito and injecting the contents into a virgin female, his mentor George Craig Jr. had shown that the virgin now acted as if she'd already mated. No longer interested in the males, she rejected their incessant advances, and went about searching for a "blood drive" to supply the meal that would allow her to make a batch of eggs. Jim was flabbergasted: A little injection had turned a flighty virgin into a calculating materfamilias. How incredible.

When he arrived as Lynn's grad student at Harvard, he tried replicating the mosquito experiments on giant silk moths—a difficult task since silk moths have scales, and Jim was violently allergic to them. Wearing a mask would reduce inhalation, but it was tough to peer down a microscope with one. Eventually the lab bought a reverse flow laminar hood in which dissections could be made with the scales carried in the other direction. But Jim's result was a bust.

In the library one day, Jim read a new article claiming that just below the esophagus inside a cockroach's head there was a bundle of nerve cells, called a ganglion, that stored a biological clock. A clock could order behavior, he told Williams excitedly when he bumped into him in the hall. "That's all wrong," Williams brushed by him, which left Jim rather crushed. That evening he decided that he had to find out why Williams thought it was all wrong, but rather than using cockroaches, he'd stick to the giant silk moth. Lynn raised these beautiful creatures on wild cherry trees in rural areas surrounding Concord. Each spring, she'd bring pupae out of diapause and let them emerge, mate, and lay eggs in large shopping bags. Lab members would gather in the early mornings and travel to the tree sites, guided by Lynn, who never needed a map to return to her favorite places. They'd pull huge nets over entire trees, place egg papers within the nets, and at the end of the season the cocoons would be harvested and separated for experiments. And they'd stop for ice cream on the way back to Cambridge.

What Jim eventually discovered was really quite amazing. Knowing that the only part of its body the moth* could move while metamorphosing was its abdomen, he secured the pupa by waxing one end of a thread to the tip of the abdomen and attaching the other end to a lever that wrote on a slowly revolving drum. That way he could look at the rhythm of abdominal movements over the day-night cycle. For three weeks he carefully recorded five pupae. And while

* Jim performed this experiment on the Chinese oak tussar moth.

they emerged as adult moths over a period of three days, all five did so *in the exact same one-hour window*, in the late afternoon. This was thrilling but what came next was truly unbelievable. When Jim removed the brain from these pupae early in development, they survived to complete metamorphosis, emerging as brainless adults—but only at completely random times. If, on the other hand, he implanted a brain into the abdomen of one of these brainless animals, they'd emerge at the appropriate time. He even implanted a brain from a different species into a brainless host only to watch it emerge as a moth precisely at the hour typical of the brain donor species, rather than the host. Incredibly, Jim had moved a biological clock from one animal species to another.

The once-in-a-lifetime event in which a moth emerges from its pupa (called *eclosion*) had to be controlled by the brain. But while performing his Wigglesworth-like experiments, Jim hadn't hooked up the nervous systems of the implant and the host, so the timing of the eclosion couldn't be a neurological event. Rather, it must depend on some secreted molecule, something like a hormone. The search led him to the discovery of a new insect hormone he named *eclosion hormone*. Incredibly, he could take an extract of this hormone, inject it into the abdomen of a moth that had finished metamorphosis and was waiting for the proper time to emerge, and get the moth to begin its "emergence sequence": a thirty-minute period of frequent rotary movements of the abdomen followed by a thirty-minute period of quiescence. Then, with repeated contractions of the abdomen in the next thirty minutes, the moth would propel itself forward and out of its old cuticle. Jim saw this same behavior when he manually removed the covering cuticle; the hormone was clearly inducing the behavior necessary for shedding the cuticle, even if there was no cuticle there to shed. Carefully opening up the abdomen of a moth to reveal the underlying chain of four nerve clumps (called ganglia), Jim connected electrodes to a number of motor nerves, added eclosion hormone, and got the same result: The complex one-and-a-half-hour program of behavior was programmed into the central nerve system. Finally, and most incredibly of all, when he cut out the ganglia from the abdomen and just placed them in a dish, the addition of the hormone generated the same ninety-minute program, a precise recapitulation of the behavior shown in the whole animal. The people who had chosen him to be a Harvard junior fellow had made a solid bet.

Meanwhile, shortly after they married, Harvard offered Lynn and Jim a very attractive deal. Sixteen miles from downtown Cambridge, in Bedford, a former military installation used to store Nike missiles had been donated to the university and converted into a field station. A new faculty member named Richard Taylor would be in charge of research; Lynn and Jim could live in the former officer's quarters, and would be given plenty of space for their own work. Taylor's vertebrate locomotion studies involved research on a number of exotic animals, and as caretakers and sole human inhabitants of the station, Jim and Lynn would be responsible for feeding and cleaning the cages of cheetahs, chimpanzees, baboons, bush babies, wallaroos, and tenrecs, alongside the more mundane dogs and goats. Taylor lived half a mile from the station and had a small corral and horse stables at the back of his property; if Jim and Lynn wanted to use it, he said, they were welcome. Rekindling Lynn's childhood love, the couple bought two horses, a happy addition to the menagerie of animals that already surrounded them. The former military installation sported two underground silos. Isolated from the temperature and light cues of the day-night cycle above, they were a perfect place for studying biological clocks.

Back at Harvard, where Jim and Lynn also had labs, an exciting new resident had recently arrived on campus. The tobacco hornworm (*Manduca sexta*) is a large brown moth that gets its name from the hornlike structure at the rear end of its gigantic caterpillar, and for its ability to sequester and secrete the neurotoxin nicotine from the tobacco plant. Easy to rear in the lab, big enough to dissect, and, most importantly, generating every six weeks as opposed to the silk moth which had only one generation a year, *Manduca* was a godsend. And since it possessed a diapause determined by recurring cycles of light and dark that could be produced artificially, it was a perfect animal for doing experiments. A student of the Cretan professor Fotis

Kafatos named Ray Hakim had been responsible for bringing it to the Harvard Biological Laboratories, but it was Lynn Riddiford who would tame the moth, making it, in her words, the "laboratory rat of insect endocrinology."

"What questions did you want to answer?" I ask her. Well, how metamorphosis really works. What makes an egg hatch a larva, a larva become a pupa, and a pupa produce an adult? Wigglesworth and Williams had laid out the basic framework of metamorphic control but there remained many holes in it: What were the genes involved? And might hormones work directly upon them? Early work out of a lab in Germany* seemed to suggest they might. But there were so many questions no one knew how to answer.

Working with *Manduca*, Lynn soon made an important discovery. When an insect grows, it molts to escape its exoskeleton, a process called *ecdysis*. The crux of this process is exchanging an old skin for a new one. Since *Manduca* produced many offspring in a short time, it also produced more mutants; one of these caterpillars was black instead of green. This opened up an amazing opportunity. By placing a small drop of juvenile hormone on the black skin at just the right time, Lynn showed the exposure would create a green spot at the next molt. Next, she exposed a piece of the skin to the molting hormone ecdysone, and implanted it on a larva about to undergo its penultimate molt. (It wasn't really hard to do, she laughs delightfully, though maybe a bit more tricky than just cutting off heads like Wigglesworth.) After it underwent its final molt, Lynn recovered the implant (easy to do, again, since thanks to the black mutant it was of a different color). What she wanted to know was whether the epidermis had turned into a larval or pupal cuticle. If it had turned into a pupal cuticle, that would mean that the exposure to the hormone had committed it to pupation even though it was on the body of a larva that was still undergoing its penultimate molt. Lo and behold, that's just what happened. Exposure to the ecdysone hormone had a dramatic

* Peter Karlson's lab in Munich. Later it was shown that hormones work on genes by first binding to nearby receptors.

effect: In the absence of juvenile hormone, it made a caterpillar larva commit to making a pupa.

Commitment. That was the crux of it. And there was a whole system of cues and clocks and molecules that brought it about. But before Lynn could figure out how the genetic control of metamorphosis worked, Harvard would need to decide whether it was committed to her. As it turned out, it wasn't. Lynn insists, even today, that it had nothing to do with her gender (Jim: "She doesn't have a feminist bone in her body!"). When she was appointed assistant professor in 1966, she says, she was told at the outset that she wouldn't get tenure since there were already too many insect people in the department. She was brought up for tenure, nonetheless, and declined, despite the fact that she already had an impressive four single-authored articles published in *Science*, and was pulling in good grant money. Whether or not her sex had anything to do with it, she and Jim would need to find a new home.

That's when they came out west, to Seattle. It wasn't yet the birthplace of Amazon and Starbucks, and the couple were able to afford a ten-acre farm, where they held three horses and filled their house with cats. Of course, Jim and Lynn brought *Manduca* with them, and began working at the University of Washington, whose faculty they joined. They decided that they would not have any kids; until and after their retirement, when they moved first to the Janelia Research Campus of the Howard Hughes Medical Institute in Virginia, and later back west to Friday Harbor island, they'd devote their lives to metamorphosis.

When we visit the Friday Harbor Marine Research Station together, I'm struck by the serenity of the place, and its gravitas. The station occupies a cluster of wooden huts on a grassy campus overlooking a San Juan Islands bay, just a short drive from Jim and Lynn's house. Narrow paths wind through groves of shade trees, past a gleaming sculpture representing the green fluorescent protein (a molecule discovered at the station). An American flag flutters high overhead. The University of Washington, to which the station belongs, kindly allows Jim and Lynn to work in their spacious lab here, for as long as they'd like. "You want to learn what we uncovered about metamorphosis?" Jim asks. "Come, let me tell you the story of the molecular trinity."

The control of insect metamorphosis had long been a mystery. As with the sea squirt and starfish, there were clearly different genetic programs running at once. If an insect undergoes complete metamorphosis, then its genome contains the instructions for making three distinct body forms: the larva, the pupa, and the adult. But what were the genes that made this system work? From the beginning, and for many decades, clue-by-clue, this would be Lynn's Holy Grail.

When they were still at Harvard, experimenting in their abandoned missile silos, Jim had shown that there was a "critical period" during the day in which *Manduca* larvae molted. This had to do with the controlled release of Williams's "brain hormone" (rechristened the master *prothoracicotropic hormone*, or PTTH, in 1972), which triggered other hormones down the road. There were three distinct phases in which the insect loosens the connection between its old and new armor, shedding its exoskeleton, expanding, and then hardening a new one. Jim also showed that there was a later "critical period" during the molt in which juvenile hormone prevented the cuticle of the larva from being colored.* All of this indicated the existence of a clock.

Meanwhile, Carroll Williams and his student Fred Nijhout proved that there was also a "critical weight" above which the larva would undergo metamorphosis on schedule, whether it was fed or starved. Internal worlds—clocks and hormones—were combining with feeding behavior to bring about metamorphosis.

---

* Other talented graduate students working on similar questions at the time were Lynn's students Mary Nijhout and Margie Fain and Carroll Williams's student (and husband of Mary) Fred Nijhout, as well as the undergraduate Lou Safranek.

In Seattle, Lynn returned to her skin-grafting experiments, the ones that clarified that after a certain point in time, *Manduca* became committed to making a pupa. The moment this happened, Jim showed, was during the animal's transition from feeding to wandering: the moment when it stopped eating tobacco leaves, coated its body with a secretion, and began building a pupa underground. This was correlated with body size, which meant that it had to do with eating. But the clocks and hormones involved suggested that the control was primarily a matter of processes that were internal to the larva: a series of spikes in PTTH release, a reduction in juvenile hormone release, and the accompanying elevation of the molting hormone ecdysone.* Hormones work by being secreted from glands into the body, reaching target cells, and then latching on to receptors on chromosomes inside those cells' nuclei that influence the expression of DNA. In this way, hormonal fluctuations could be sensitive both to external environmental cues and to internal clocks, bringing about metamorphosis. But who were the true masters of this affair? Wigglesworth and Williams didn't know—no one knew—and Lynn would have to gain new skills to find out.

Genetics had come a long way since Mendel. In the first half of the twentieth century, a Kentuckian named Thomas Hunt Morgan had turned his lab at Columbia University into the "Fly Room," where his team of pioneers fashioned *Drosophila*—the fruit fly—into the leading model organism of the field. Mendel's long sought after elements turned out to be actual material things, within the nuclei of cells on chromosomes, and they built and maintained bodies by making proteins. Just around the time when Lynn was falling in love with science as a teenager at her Maine summer camp, James Watson and Francis Crick at Cambridge University, and Rosalind Franklin and Maurice Wilkins at King's College in London, were figuring out the structure of DNA. That same year, 1962, when Lynn was writing her PhD at Harvard, François Jacob and Jacques Monod in Paris showed

* To be precise, the active form of the ecdysone hormone, called *20-hydroxy-ecdysone*, or 20E for short.

how genes are actually turned on and off: Proteins upstream from them on the chromosome bind to receptors that regulate them, and the genes themselves regulate the proteins in turn. That way, when a cell needs to produce a certain protein—for example, in order to make a hormone—it can turn on the right gene, and end production when appropriate. Neither heat nor energy nor a spirit were responsible; in the end the secret of heredity came down to a molecular code. A cell's machinery could read, replicate, translate, and pass on the code to the next generation. Spemann and Mangold had been right: Building bodies was about control and regulation. Now molecular biologists would need to do the hard work of finding out which genes controlled what traits and how.

Lynn's generation was the first to step up to the massive challenge. And over time it became apparent that what genes really are and how they work is more complicated than the pioneers of molecular biology first thought. Still, over five decades—including sabbaticals to gain new tools and with the help of colleagues*—Lynn meticulously put the pieces of the metamorphosis puzzle together. It was especially crucial that the lab switched from working on *Manduca* to the *Drosophila* fly, which had been well characterized as a biological system since the early days in Morgan's "Fly Room." There were already nearly two million lines of *Drosophila* available, with tools to knock almost any of its genes out, or alter their expression, as Lynn wished.

It was toward the end of the 1990s that Lynn's journey took its first dramatic turn. That's when she finally found a protein that turns on the gene responsible for an insect's transformation from a larva to a pupa. Since the protein controls the rate of transcribing information from DNA to messenger RNA (a molecule that helps translate the information in the DNA into a protein) it is called a transcription

---

* In particular sabbaticals at Bob Shimke's lab at Stanford to learn more molecular biology, and at Michael Ashburner's lab in Cambridge, UK, to learn more fly genetics, and the help of Xiaofeng Zhou and Kiyoshi Hiruma in Lynn's own lab.

factor, and this particular transcription factor was called *broad*. Lynn was never able to find the elusive juvenile hormone receptor that got this whole cascade rolling (two former post-docs eventually did),* but a researcher named David Martín working in collaboration with Xavier Bellés in Barcelona showed that a second transcription factor, *ecdysone-inducible protein* or *E93* for short, controlled the gene responsible for turning a pupa into an adult, in cockroaches, beetles, and flies. Step-by-hard-earned-step, the gene hunters were inching closer to the holy grail.

As it turned out, a forgotten paper authored by Carroll Williams and Fotis Kafatos in 1971 had postulated three master genes that could be responsible for the three stages—larva/pupa/adult. In the very same genome, the two men had speculated, three separate developmental controls could have evolved to influence radically different traits in one part of the life cycle without influencing traits in another. A researcher named Nancy Moran later wrote about how such "adaptive decoupling" could explain complex life cycles. And while Lynn and Jim remained oblivious to the Moran paper from 1994, and the field had completely forgotten the Williams-Kafatos paper from 1971, the search continued. It was two down, one to go.

Soon a red herring presented itself. The small community of metamorphosis experts became convinced that the third master gene was a gene called *Krüppel homolog 1*, but Jim and Lynn were skeptical. When a molecular cousin of *broad*, which was known to play a role in determining the different types of adult fly brain cells, presented itself, the penny dropped. The cousin was *chinmo* (so called since it induced *ch*ronologically *in*appropriate *mo*rphogenesis in an embryo), and like many other genes, it had more than just one role. What Jim and Lynn figured out was that if the nervous system transitions from *chinmo* to *broad* to get brain development right, maybe it makes the same switch to move from larva to pupa. It was a good idea

* Marek Jindra and Jean-Philippe Charles showed that the juvenile hormone receptor is *Methoprene-tolerant* (*Met*). The *Met* gene was first identified by T. G. Wilson and J. Fabian in 1986.

Jim and Lynn with moths in the late 1990s. Courtesy of Jim Truman.

but seemed impossible to prove: Repressing the expression of *chinmo* leads to death of the embryo, so how would one uncover its role in metamorphosis? To get around this problem, Lynn and Jim found a brilliant solution: By removing *chinmo* from just the front, and not the back, halves of larval segments and imaginal discs, they could compare cells that lacked it with their normal neighbors.

Just a year after the appearance of Martín's discovery of *E93*, at seventy-seven and eighty-six years of age respectively, Jim and Lynn published an elegant paper in the *Proceedings of the National Academy of Sciences*. Considering the chain going back through Wigglesworth to Swammerdam to Harvey and all the way to Aristotle, their accomplishment was nothing short of astonishing: the mystery's long-awaited crowning moment, the holy grail, brought about by two small-figured people observing the smallest changes in tiny insects with the greatest amount of care. Three master genes—*chinmo, broad, E93*—orchestrated insect metamorphosis. Jim and Lynn called it "the molecular trinity," drank a nonalcoholic beer and an orange juice, and took a moment to rest.

Of course, metamorphosis itself was less tidy than the trinity suggested—more like a mosaic than a three-colored flag. Over a long career of experimenting, Lynn and Jim had learned the hard way that the textbook explanation of how juvenile hormone controls metamorphosis—the slogan "Hi-Low-No"—was way too simplistic. In fact, juvenile hormone weaves in and out during the life cycle of a butterfly or moth, and its expression looks like a tapestry. Eyes and wings, legs and abdomens, guts and brains, even different segments of the skin, each mature at different rates. Yes, the general picture is that when juvenile hormone levels are high and *chinmo* is expressed, *broad* and *E93* are turned off.* Yes, when PTTH spikes it turns off juvenile hormone and turns on ecdysone, signaling to *broad* to start the program to pupate. Yes, when pupation is complete, *E93* kicks in and shuts down *broad*, to finally produce an adult. But take the eye, just as one example: While juvenile hormone needs to subside to bring about pupation, if you don't bring it back for a bit just before the insect sheds its outer cuticle, instead of developing a small pupa eye the pupa will develop an outsized adult eye, like a grotesque mythological monster.**

Lynn had spent a career showing that juvenile hormone was the "status quo" hormone of insect metamorphosis, but more than its *concentration*, it was the careful tweaking of the *timing* of its action that mattered. As an insect develops, juvenile hormone action has to be channeled in such a way that the balance is tipped to a larval, pupal, and adult framework, each under the control of its master

---

* In a recent meeting of entomologists in Kyoto, Lynn and Jim's longtime friend and colleague Xavier Bellés presented a fascinating new find: In the cockroach *Blattella germanica*, *chinmo* represses *E93*, but it does not repress *broad*. The embryonic role of *broad* is still poorly understood in hemimetabolous insects, but a change in its relationship to *broad*, namely, preventing *broad* expression in the embryo, could be the key that opened the door to making a true larval stage.

** Likewise, without juvenile hormone coming back *after* pupation, adult flies fail to show any mating behavior, and don't produce the major sex pheromones, besides.

gene. This tipping from one stable state to another brings about a chain reaction. In professional lingo, downstream actions of these master genes turn on subregulatory gene networks, which themselves turn on unique larval/pupal/adult gene sets below them. It's a delicate, intricate, exquisitely beautiful developmental symphony, but the symphony can never be played backward: Once an insect becomes committed to either being a larva or a pupa or an adult, the commitment is irreversible. Growing up is a one-way affair.

How, when all is said and done, does a gooey caterpillar morph into a divine butterfly? From the very beginning, directed by genes in the organizer in the egg, cells from the endoderm, mesoderm, and ectoderm begin to build the body of a larva. But other cells, destined to produce most of the adult body, are set aside as imaginal discs—the structures Swammerdam showed a gaping audience at Melchisédech Thévenot's home back in the seventeenth century in Paris.* When the time is right, ecdysone is secreted and juvenile hormone subsides to awaken those imaginal disc tissues from their slumber. Spemann and Mangold hadn't found the actual levers in their organizer, but they did discover a general principle: In embryonic development, regulation is key. The same holds true for post-embryonic development. When it comes to fashioning creatures, it's not God or the will of the universe that's responsible. Simple molecules made from Earth's organic elements—genes carefully modulated by hormones—are nature's willing accomplice.

I look at them from across the lab. Jim and Lynn are not flamboyant like Haeckel. They're not tortured, like Merian. In fact, they're the opposite of their predecessors, gleeful and uncomplicated. They lost their religious faith gradually, they tell me; the deeper they drilled down into the workings of metamorphosis, the more the appeal to God seemed unnecessary. But belief wasn't replaced for them by

* Now we know that Swammerdam was lucky: Only in butterflies and moths do the imaginal discs actually resemble wings and head with proboscis—in flies they appear as blobs. Had Swammerdam chosen a fly, his trick would have flopped.

philosophy. Like Merian and Haeckel, Jim and Lynn are tenacious seekers of truth, but they don't overthink things.

"What's a self to me?" Lynn smilingly repeats my question, having devoted her life to unraveling the molecular minutia responsible. What brings a self into being?

"I don't know," she laughs. "I've never actually given it a thought."

7.

HOPEFUL MONSTERS

For forty-two years he'd worked without incident, peacefully examining tiny sea creatures. Now he was retired but would come every day to what was lovingly called the "Geriatric Wing." It would be hard to say there was much action up at the Port Erin Marine Laboratory on the Isle of Man, midway between the waters of Blackpool and Belfast, where on most days the sun hid bashfully behind foreboding clouds. But while seeking out marine invertebrates one morning in the spring of 1990, the man with the laughing eyes slipped and hit his head on a wet rock. It was a silly thing for a veteran planktologist to do. Having driven himself to the hospital, he was now in the throes of a stroke, able to hear but unable to talk. It was rather peaceful and quiet in this liminal state, he would later report. An opportunity to collect his thoughts.

Here's what he believed: Darwin was wrong. All that gradual step-by-step evolution was an illusion. Or, perhaps not an illusion entirely—change does proceed gradually, but this was hardly the entire story. Alongside the slow, measured climb, the plodding grind, nature acts in dramatic fashion. She moves by leaps and bounds. And so, despite ushering in a revolution, deep down Darwin was conservative. He placed *Natura non saltum facit* at the heart of his theory of evolution: Nature does not make jumps. But she does. In a big way. Donald I. Williamson was sure of it.

Williamson had grown up near the coast of Northumberland, England, exploring the tidal offerings with his father, a schoolteacher and amateur naturalist. Completing a degree in zoology in 1942, he left the university to join the Royal Navy, becoming a radio operator in the Mediterranean. On board HMS *Antwerp*, the floating headquarters for the Allied invasion of Sicily, he contracted tuberculosis and was invalided to a sanitorium in England. His illness dashed his dreams of becoming a teacher, like his father. But at the sanitorium, Williamson came across Darwin's *Origin of Species*, and a disability allowance now allowed him to pursue a doctoral degree. He wrote his dissertation on tiny beach-dwelling crustaceans called sand hoppers, and in 1948, when the University of Liverpool needed a planktologist on the Isle of Man, even though he knew very little about plankton, Williamson took the job.

As he began to study invertebrate larvae in the Irish Sea, he soon discovered that many such plankton had not yet been described. It was striking how many shapes and forms there were, their permutations both distinct and endless. If you didn't know better, it could almost make you believe in a God. Gradually filling holes in the literature over many years of careful observation, Williamson became a world expert. And the more he learned, the more a suspicion grew within him: Perhaps Darwin's *Origin of Species* had been all wrong.

Take the metamorphosis of starfish. If evolution proceeded piecemeal, why were there two separate developmental programs, one for the juvenile and another for the adult? It seemed so unnecessarily wasteful, so disjointed and bizarre. Similar questions arose when considering the sea squirt. A chordate juvenile sacrificing itself to become an adult invertebrate: How could that make sense if natural selection worked gradually, as Darwin thought? No less striking was the fact that many larvae of crustacean species looked nothing at all like each other. Instead they closely resembled the larvae of creatures far removed from them on the evolutionary tree. It was as if the adult crustaceans had jumped from a branch on one side of the tree and grafted themselves onto far-flung juvenile branches on another. The chimeras of ancient Greek mythology described many ocean life-forms better than the bearded English prophet. Evolution wasn't the linear process many had been led to believe.

One of Darwin's younger contemporaries had made similar observations. Francis Maitland Balfour was not drawn to politics like his older brother Arthur, who would one day become prime minister. Instead, Francis became his generation's leading animal morphologist; Darwin himself called him the "English Cuvier."* Working at the University of Cambridge in the 1870s, Balfour came to challenge Ernst Haeckel's and Darwin's view of evolution and animal development. He died in the summer of 1882 attempting to climb the yet unscaled Aiguille Blanche, Mont Blanc, in the Italian Alps. Just thirty-one years old, he was widely considered the greatest loss to science of his age.

What had Balfour seen? The same evidence as Williamson: that many marine larvae look nothing at all like their adult incarnations. And he doubted very much that such larvae represented embryonic "fossils" of the animal's evolutionary past. Yes, humans have gill slits and tails as embryos, but gill slits and tails are features of the marine vertebrates, our cousins. If starfish and their larvae, on the other hand, were all likewise from one lineage, it made little sense that they had completely different body plans, not to mention the reversion from chordate to invertebrate, as performed by the sea squirt. Rather than the adult ancestor of these species resembling their current larvae, as Haeckel would have it on his law of recapitulation, Balfour suggested that the larvae had somehow been "introduced into the

* You'll recall, George Cuvier (1769–1832) was a French zoologist and anatomist, often referred to as the founding father of paleontology, and a major figure in the natural sciences in the nineteenth century.

ontogeny of the species, the young of which were originally hatched with all the characters of the adult." In other words, current larvae of starfish and of sea squirts are "secondary" larvae of ancestors who went through metamorphosis long ago, but whose original, or "primary" larvae, had been lost to time and natural selection.* Sometimes, species are shaped by dramatic interventions, like the arrival of a foreign larval stage. But how precisely did such dramatic interventions occur? How could one species literally kidnap the larva of another species, and make it a part of its own development? Sadly, Balfour took his thoughts to his alpine grave.

Though Williamson was not aware at first of Balfour's work, he inherited his questions. When Williamson drew a phylogenetic tree of marine crustaceans based on their larval forms, and when he compared it to the tree based on adults' forms, he could see very clearly that the two did not overlap. The easiest explanation was that the larvae of widely separated species looked alike because of what's called "convergent" evolution, the process by which a whale and a fish, for example, have developed similar tails and fins and streamlined bodies despite sharing no common ancestor. Williamson rejected this idea: Fish and whales had adapted to a similar environmental challenge. Invertebrate larvae, on the other hand, were incredibly diverse in form despite inhabiting the same exact ecological niche. Those larvae that happened to be similar in structure were therefore not the result of convergence. It was more logical, he thought, that larvae originated elsewhere first, as adults, and then somehow became the juvenile stage of a different creature's adults. On this scheme, two separate genomes would have combined in one species, the first genome conducting the development of the larva and the second conducting the development of the adult. If this was true, then not all evolution was vertical, with each generation of a species growing out

* Balfour suggested, too, that the octopus and squid were examples of originally metamorphosing species who had lost their primary larvae and never regained new ones, evolving to develop directly into miniature adults from their fertilized eggs instead.

of the one that came before. Instead, radically different species were occasionally spawning with each other, and evolution was—at least sometimes—horizontal.

He sat down to think things through. A species is shaped by natural selection, but there's a great role played by chance. The seeds of plants, for example, are carried by winds to unknown fortunes. Many crustaceans and cephalopods and fish, like crabs and squid and marlins, reproduce with a messy hit-or-miss technique called "broadcast spawning," fertilizing their eggs outside their own bodies. Was it so crazy, considering all this, to imagine a flying insect, deep in the evolutionary past, from whose oversexed abdomen a tiny droplet of sperm dropped from the heavens one day? Might that sperm have fallen directly onto the egg of a worm in the undergrowth, and by good fortune managed to invade it? With the numbers of such creatures in the quadrillions, and in the grace of almost infinite time, might chance not have awoken this egg, just once, and produced an offspring? After all, different species can produce offspring: Female horses and male donkeys make mules, and zebras and horses make zorses. Neither zorses nor mules can procreate, Williamson knew, but what about ligers (the offspring of male lions and female tigers), or tigons (made from female lions and male tigers)? They're both perfectly fertile. Granted, these creatures are similar, but remember: For eons life evolved underwater. In the great expanse of the oceans, where the currents act as the winds do for plants on land, all kinds of wild unions might be possible. It actually made sense that for marine invertebrates, "hopeful monsters" might be closer to the rule.

To test his hypothesis, Williamson fashioned himself a matchmaker: Whatever others would think of him, he'd try to mate a sea urchin with a shrimp. This was not a matter of mating a horse with a zebra, or of orchestrating the kind of match that often occurs in the wild between a cackling goose and a barnacle goose. Williamson was proposing to mate two species of different phyla with completely different body plans—a bilateral crustacean with a radial echinoderm. A gulf of 540 million years separated them from their common ancestor. But bringing two genomes together in one offspring was the only way to explain anomalies like the starfish and the sea squirt. Or so he thought.

Down to Scarlett Point on the Irish Sea Williamson treaded with his daughter, nets in hand, until they caught a sufficient amount of female specimens of the shrimp *Gammarus duebeni* and brought them back to Port Erin. There, at the Marine Laboratory, he'd already milked a good number of the male sea urchin *Echinus esculentus* of their semen. The bright pink prickly sea urchins were difficult to sex, but all he needed to do is flip them over a beaker; they'd either lay their eggs or squirt out their sperm, and Williamson soon became a pro. *Gammarus* shrimp do not metamorphose, developing instead from egg to adult directly. Perhaps if he bathed the female shrimp's eggs in a pool of sea urchin semen, they'd acquire the genetic program to make a larva in between. Williamson dubbed this his "theory of larval transfer." And, carefully placing the spawning female shrimp into a concentrated cloud of sea urchin semen, he held his breath.

It took a while to learn how to pry the shrimp's eggs from beneath their privates with tiny needles, but Williamson mastered the delicate movements, and rushed to place their eggs beneath a microscope. He peered down the ocular for a sign of life, day after day, but *Gammarus* eggs being quite opaque, saw little at first of any kind. Most of the shrimp eggs died within a day or two, and those that survived longer exhibited no signs of development. But in two tiny eggs out of the hundreds, something seemed to be different. Spheroids moved about, spinning rapidly propelled by cilia. Unfortunately, neither of the eggs hatched, and when Williamson dissected them whatever was inside disintegrated. But here was the catch: Shrimp like *Gammarus* possess no cilia, at any stage. And the larvae of sea urchins like *Echinus* begin life as tiny balls of ciliated cells. On the Isle of Man, Williamson didn't have the photographic equipment necessary to see clearly what had developed within those eggs, and he'd produced no live larvae. But he was sure he was on to something, at last.

So he dreamed up another Ovid-like experiment. This time he would marry a maternal sea squirt to a paternal sea urchin—a urochordate to an echinoderm—again two creatures separated by an evolutionary chasm, and according to orthodoxy not likely to produce an offspring of any kind. There was a danger: The sea squirt was a hermaphrodite, making both eggs and semen that could self-fertilize, a fact that might foil his plan. And so after he milked the males, Williamson waited for a good number of hours, and threw aside the eggs that showed any sign of development. The remaining eggs he placed in a dish, dousing them with sea urchin semen. He knew that normal sea squirt larvae turn into tadpoles, and that normal sea urchin larvae begin as balls of ciliated cells before elongating into prism-like bodies called pluteus, with eight little arms stiffened around a mouth and an anus. Then, the adult form starts developing within.

Lo and behold, in the winter of 1987, two eggs hatched, and out came two spherical larvae. As they grew, they elongated, and then out sprouted the eight arms. He'd done it! Williamson had created a sea urchin pluteus from a sea squirt egg! Even though the chimeric larvae did not develop from the pluteus to any form of adult, he was thrilled, dumbfounded. If a silly planktologist from Liverpool had succeeded in just a few lucky experiments, he figured, one could only imagine what the vast ocean could do in the expanse of time. Williamson was convinced: Larval transfer was real.

Quickly, he wrote up a paper and sent it off, first to the *Biological Journal of the Linnean Society*, then the *Oxford Surveys in Evolutionary Biology*. Next came the *Journal of Natural History*, the *Journal of Zoology*, and *Biological Reviews*. All rejected him. His result just seemed too out of this world. Finally, a brave editor of a journal called *Progress in Oceanography* took a chance on him, writing in his preamble that based on the rejections he'd received, "Darwin would have probably had less trouble submitting a draft of the *Origin of Species* to the Bishop of Oxford." Williamson took a chance, too, and sent a reprint across the Atlantic: Lynn Margulis was an iconoclastic biologist who had once suggested that complex life came about on the planet thanks to single cells from one lineage swallowing up smaller ones from another lineage billions of years ago. Though younger than him, to Williamson Margulis seemed a kind of intellectual parent. People had ridiculed her, calling her a crackpot and an anti-Darwinian. But her theory of endosymbiosis was ultimately accepted.

Margulis wrote back immediately: "Who are you? I've never heard of you. What a marvelous theory!" to which he replied: "There's no reason why you should have heard of me. I'm not an evolutionary biologist. . . . I am sixty-six years old, from a family whose members are short-lived and thus I'm on a straight-line course for posthumous recognition."

Margulis moved quickly. Following further rejections from *Science* and *Nature* of a new paper detailing Williamson's ideas, she suggested that he send a shortened version to the *Proceedings of the National Academy of Sciences*, where she was by now a respected member. The replies came hard and fast: One reviewer refused point blank to even read it; "High school level rubbish," wrote another; a third kindly urged his colleague to distance herself from Williamson lest her own reputation be irrevocably tarnished. The plan had backfired. Badly.

A boundary had been overstepped, quite literally, and the community wasn't amused. How are species made? To most biologists the answer was obvious: Boundaries develop over time, either geographical or physical or behavioral, which do not allow certain individuals in a species to mate. Over time, as these boundaries drive one "family" further apart from another, slight variations grow into full-blown species. This theory is called the Biological Species Concept, and the capital letters suggest its importance. Species are real categories, not to be tampered with. Williamson's theory of larval transfer said otherwise.

Despite the fact that Williamson had already produced hundreds of fertilized eggs and pluteus larvae from the eggs of sea squirts, he was shouted down at conferences. "This is impossible!" one expert spouted from the audience incredulously, unwilling to accept that nature had produced chimeras long before genetic engineering, not to mention Roman myths. But Williamson didn't give up. In fact, he began to believe that he was making serious headway. Continually refining his technique, his hybridization experiments now produced many thousands of metamorphosing hatchlings, and even a few viable adult sea urchins that had come from the pluteus emerging from the semen-dowsed sea squirt egg. To his credit, the shouter at the conference (Dick Whittaker, an expert on marine ascidians, and the same reviewer of Williamson's PNAS paper who had called it "high school level rubbish") invited him to his lab in America to perform a larval transfer experiment. No viable larva hatched, but the shouter could see that some eggs were dividing. If this were possible, perhaps what Williamson had been telling the world was more than pure nonsense. Perhaps non-Darwinian evolution was real.

It was right about then that Williamson slipped on the rocks and hit his head, not far from Point Erin, leaving him, literally, speechless.

Postscript

Donald I. Williamson eventually overcame his stroke in the early 1990s, no small feat since in his case aphasia (the inability to speak) was accompanied by dysgraphia (the inability to read or write). Paralyzed in his right hand, he taught himself to type with his left. And he made a new friend. Robert was a seventeen-year-old patient who shared a room with him at Noble's Hospital on the Isle of Man. Struggling to explain his ideas to him, Williamson wasn't giving up on his theory.

Following a long series of rejections of a book he'd completed before the accident, and with the help of Margulis and others, the New York publisher Chapman and Hall finally printed *Larvae and Evolution: Toward a New Zoology*. It was 1992, the year of his seventieth birthday. Williamson took pains to stress that he was a fan of Charles Darwin's. That evolution was true, and that natural selection was its main motor. It was just that the kind of universal, gradual, linear speciation advanced by Darwin was often challenged by the facts. Williamson was sure of it. Now the rest of the world needed to come around.

Alas, the attempt to downplay his heresy did little to protect him. Molecular genetics was revolutionizing evo-devo, opening up the black box between genotype and phenotype, taking out what was found in the box and then figuring out how to put it back to make evolutionary sense. If branches far removed on the tree of life sometimes fused and came together; if evolution could be reticulate rather than always linear—well, if that was true, very disparate sets of genes would need to somehow come together and learn to cooperate. Genetic evidence of fusion, in other words, would be the ultimate test of "larval transfer." Until someone came up with the evidence, Williamson would continue to be viewed as a crank.

Finally, the crushing blow came. When a bona fide geneticist stepped in to set things straight, the house of cards built by the little-known planktologist from Liverpool came crashing down. Michael Hart, from the Institute of Molecular Biology and Biochemistry at Simon Fraser University in British Columbia, requested and was kindly sent samples from three of Williamson's sea urchins. Williamson claimed they had developed from the paternal pattern of pluteus larvae which had sprung from the maternal sea squirt egg. If this was true, they would surely exhibit some sea squirt genes. But when Hart carefully examined the DNA of the sea urchins he discovered an uncomfortable truth: There was no sea squirt DNA to be found anywhere. Instead, it seemed as though the putative hybrids were merely the result of a sea urchin egg fertilized by sea urchin sperm. Both men knew that some sea urchins are hermaphroditic, and could have fertilized their own eggs. Despite his credentials as an old-school naturalist, Williamson had simply failed to notice.

Williamson denied it: Sea urchin and sea squirt eggs look nothing alike, there's just no way he could have been so sloppy. Besides, a colleague at Port Erin had shown that only one in three thousand *Echinus esculentus* sea urchins were hermaphrodite, making hermaphroditism an unlikely explanation of his results. But Hart insisted: "It would have been an enormous boost to my career (not to mention Don's hypothesis) if I had found tunicate DNA in his hybrid sample," he wrote to the author and biologist Frank Ryan. "Far from being disposed against his ideas, I was in fact rather disappointed at the bland predictability of my results."

A few scientists stood by Williamson and his theory. Margulis insisted that its dismissal was just another case of disruptive ideas being resisted by an ossified, threatened establishment. "We will win one way or another, because this is science," she told a reporter for *Nature*. She died just two years later, while Williamson lived to be ninety-four, despite his short-lived ancestors. Shortly before his death, in 2016, Lynn and Jim met him at the Linnean Society in London. Jim tried to explain to him why his theory was wrong, but it wasn't a fair conversation: Williamson was weak and in a wheelchair, and, as Jim would tell me, it really was no kind of a match. Williamson's supporters called for researchers to try and repeat Williamson's experiments. Others thought it a waste of time, or worse, charlatanry. Williamson had not been silenced, they claimed, he'd been amply published in the most visible forums, and the harsh verdict had been handed down. The throngs of graduate students eager to make their name by proving Williamson right never materialized. No one has come forward to claim they've been able to show that hybridizations between creatures in different phyla, as opposed to species in the same genus, are possible. As Jim and Lynn would discover, there is a fascinating story to tell about the origin of metamorphosis, which hearkens back to the oceans. But "hopeful monsters" remain a part of mythology. In nature, at least for now, they're dead.

8.

ORIGIN

The first ancestors of the insects lived in the sea. For nearly one hundred million years, they lived nowhere else. The seas were where all the action was, for the lands had yet to grow flowers, trees, or grass. In those days, only algae, liverworts, and mosses hugged the desolate landscape, a single continent later named Pangaea.

We may never know why the first aquatic arthropods climbed onto land 450 million years ago, but what is certain is that none of them had wings. For many millions of years to come, they would scamper about the waterfront, crawling into damp nooks and crannies, struggling to survive. Most likely they looked like segmented worms. Over time, their front segments fused to make heads, and nerves in the heads clumped together to create little brains. Gradually, on land, the head parts evolved into mouthparts and eyes and antennae, helping the creature to sense its new environment. Legs grew out of the appendages, and, at the back end, tails were modified for mating. The proto-insect was gaining ground on Earth, but it was made of soft tissue with no real protection. Some became the forerunners of scorpions and spiders. Others retreated back into the waters, becoming the ancestors of lobsters, shrimps, and crabs.

On land, gradually, the invaders developed a tough outer armor to make up for their vulnerability. It was a marvel of engineering: Made from the conversion of carbohydrates into a polymer called chitin, the enveloping layer covered almost all the insect's skin. It was strengthened by a protein—sclerotin—that tanned in sunlight and further stiffened the exoskeleton by banding together its sulfur groups. This is what gave the outer cuticle its horny sheen and toughness, a trick that would later be used by reptiles for scales, birds for feathers, and mammals for hoofs. But the cuticle, unfortunately, also presented a problem: How were insects to grow? With no internal skeleton powered by muscles, the coat of armor was like a prison.

Hormones were "recruited" to solve the problem, and thanks to their orchestrated action a dramatic new trick was born. With the help of ecdysone, juvenile hormone, and PTTH, as Lynn and Jim would one day find, the intricate machinations that allowed for detachment of the old cuticle and construction of a larger one were gradually put in place. In four orders of primitive insects that do not metamorphose, the molt was invented. Primitive silverfish, springtails, and firebrats crisscrossed the land.

Except that the skies loomed above, irresistible like the call of sirens. And over time, and many selections, those insects able to marshal it would be at an advantage. Gradually, a form of incomplete metamorphosis now came into existence. In the ancestors of cockroaches and termites and mantids and locusts and crickets and dragonflies, underdeveloped wing pads appeared that opened fully on the final molt.* Becoming the phylum of mandibulates (named for their side-mounted jaws), the flying insects proliferated. No surprise: With wings on their backs to escape predators and find greener pastures, the sky really did become their limit.

But the *really* big burst of life came when an entirely new type of creature evolved: The kind whose young look nothing at all like their adults. According to the fossil record, this happened 280 million

* Mayflies are an exception, growing wings to escape the water in their penultimate molt (with the final molt reserved exclusively for genitalia).

years ago, but these new species only began to explode when flowering plants arrived, 150 million years later. The more nectar the plants produced, the more they diversified to disperse their pollen. Ants, bees, beetles, termites, lacewings, butterflies, and wasps abound, growing to account for over 60 percent of all insects. This was no small number, seeing as creatures with six legs made up 80 percent of all species on Earth. The coevolution of plants and insects made complete metamorphosis a stunning success, and like all evolutionary successes, it began with an accident—in this case, when a creature hatched from its egg just a bit too early.

At least this was the idea of the Paduan entomologist Antonio Berlese. The larva, Berlese wrote in 1913, was an embryo that had hatched prematurely. Berlese's idea was disseminated by his British colleague Augustus Daniel Imms. In fact, the Berlese-Imms theory, as it became known, had a long lineage, going back from Darwin's contemporary John Lubbock, to the seventeenth-century physician William Harvey, who thought that certain insects have so few nutrients in their eggs that their embryos were forced to hatch before completing development. And Harvey, *mutatis mutandis*, was channeling Aristotle: When imperfect caterpillars grow big enough, they can finally make a pupa—the perfect, adult-bearing egg.

But was this true? Wigglesworth had shown clearly that creatures like moths and butterflies lived on a continuum with cicadas and mayflies. The latter never produced larvae and pupae like the former, but the same hormones existed in both. And so toward the mid-twentieth century, another English scientist, Howard Hinton, announced that Imms and Berlese and Lubbock and Harvey and Aristotle had all been wrong. The larva was not an embryo that hatched prematurely. Instead, Hinton argued that the larvae of completely metamorphosing creatures like butterflies were exactly equivalent to the nymphs of incompletely metamorphosing creatures like cicadas. That would mean that the pupa of the former corresponded to the final wing- and genital-bearing molt of the latter. Hinton was known for saying that "a species is good if a competent taxonomist says it is"; few were as sure of themselves as he was. And what needed explaining for him was not the larva, but the pupa: Why had the pupa evolved to create a stage of such high drama? Why had chrysalises and cocoons appeared, providing caterpillars havens to morph into butterflies and moths? The answer he gave was really quite pedestrian. Concentrating all their shapeshifting during pupation, burrowing insects wouldn't be encumbered by wing pads, and would therefore fare better in soil.*

For years the Hinton hypothesis and the Berlese-Imms theory vied for supremacy. Then Lynn and Jim came along. I ask them how it happened. After all, neither had any formal training in evolution and they'd always asked *how* metamorphosis works rather than *why* it came into being. Jim explained that when they went to meetings of endocrinologists, they met experts on fish and amphibians and birds and crustaceans, each in their own world. Evolution was always the most exciting idea that linked everyone together. Maybe looking at different creatures' embryos, just as Haeckel had, could shed light on the origins of metamorphosis. In the summer of 1993–1994, Jim and Lynn traveled to Australia to find out.

Jim began by looking at the embryos of locusts. Locusts undergo incomplete metamorphosis, but within a short time the parallels he found with complete metamorphic development in flies blew his socks off. Running to Lynn's lab just a short way down the road, the two began studying the role of juvenile hormone in the locust embryo. For Lynn this was coming full circle: Her early interest at Harvard in the 1960s had been about what juvenile hormone does *before* a moth is born, not *after*. Returning to Canberra three years later, she and Jim exposed a locust embryo prematurely to juvenile hormone, and the result was striking: The embryo stopped growing, and its cells started specializing (a process called *differentiation*)—precisely the two changes necessary for making a larva. But, of course, locusts never made true larva, only nymphs that were miniature adults, minus wings and genitalia. A thought immediately popped into their minds: Hatching from an egg

* Later, the evolutionary biologist William D. Hamilton argued that insects without wing pads would do better hiding behind bark, as well.

seems like a big deal—there's life before hatching and life after—but if you follow molecules rather than forms, maybe it's an arbitrary divide. There were three stages to incompletely metamorphosing insects: a pronymph (before hatching), a nymph, and an adult. And those three stages corresponded exactly to the three stages of completely metamorphosing insects (all after hatching): larva, pupa, adult. With just a shift in timing—what Haeckel had called heterochrony—the embryonic pronymph could have turned into a free-living larva. There was circumstantial evidence to support this: In butterflies, only part of the yolk is enclosed by the developing embryo, and developing caterpillars feed on the rest before they hatch. If the embryos of incompletely metamorphosing creatures had also failed to enclose all their yolk, it made sound adaptive sense to modify the pronymph so that it could eat while still in the egg. And if such an eating creature—a protolarva—just happened to hatch prematurely, well, in essence it would be a caterpillar.

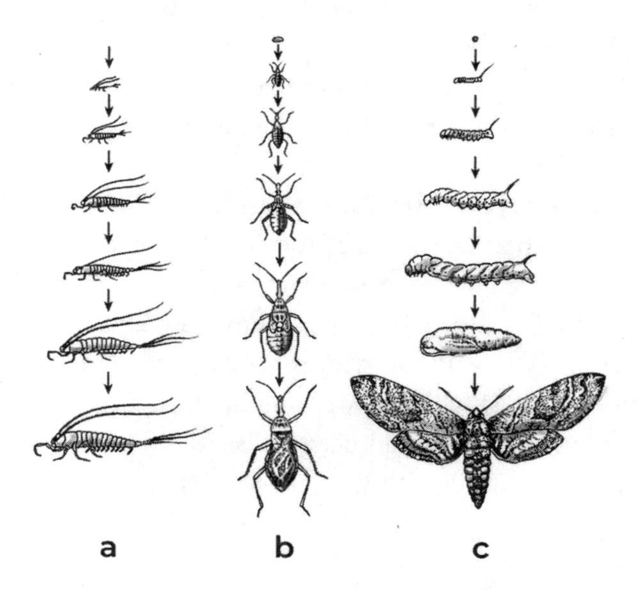

Different developmental strategies born in evolution: (a) ametaboly, in which the immature creature just keeps growing; (b) hemimetaboly, in which the immature nymph or naiad looks like a small version of the adult and gradually develops wing pads and sexual organs on its last molt; and (c) holometaboly, or complete metamorphosis, with its three distinct stages: larva, pupa, and adult. Credit: Inna Gertsberg.

If this were true, the hormones would provide the answer. Lynn and Jim wrote up "The Origins of Insect Metamorphosis" and sent it to *Nature*. Berlese and Imms were right, it claimed, the cocksure Hinton wrong.

In the meantime, in the greater world of biology, the science of evo-devo was coming into its own. The evolutionary synthesis had torn apart ontogeny from phylogeny, treating genes as abstractions that explain the evolution of populations while putting individual development into a black box. But beginning in the late 1970s, a group of scientists, sprinkled in labs around the world,* began discovering something amazing. There was a set of genes called Hox genes that specified segments of the body plan of an embryo along the head-tail axis of the animal. Those very genes, amazingly, were strung along chromosomes in the exact same order of the body parts they specified: the gene for the hind part of the abdomen beside the gene for its front part beside the gene for the thorax beside the one for the head. Even more stunningly, the very same genes specified the very same body parts both in insects and vertebrates. There were deeper ties between the creatures on Earth than anyone knew or had dared to imagine.

What came next was the imprimatur of modern, molecular science. Scientists soon learned that they could play around with Hox genes to place a leg on a head, or a wing on an abdomen. They could even plant an eye-making Hox gene from a mouse into a fly and get a fly eye despite the fact that fly eyes are compound devices whereas mouse eyes have a single lens focusing light on a retina. Hox genes were interchangeable and had been conserved for hundreds of millions of years. No wonder William Bateson, the man who coined the

* These included the trio Ed Lewis (Caltech), Christiane Nüsslein-Volhard (Heidelberg), and Eric Wieschaus (Heidelberg and Princeton), who won the 1995 Nobel Prize for their work, but also, in the 1980s, Michael Levine, Ernst Hafen, and William McGinnis working in Walter Gehring's lab in Basel, and Matthew Scott and Amy Weiner working in Thomas Kaufman's lab at Indiana.

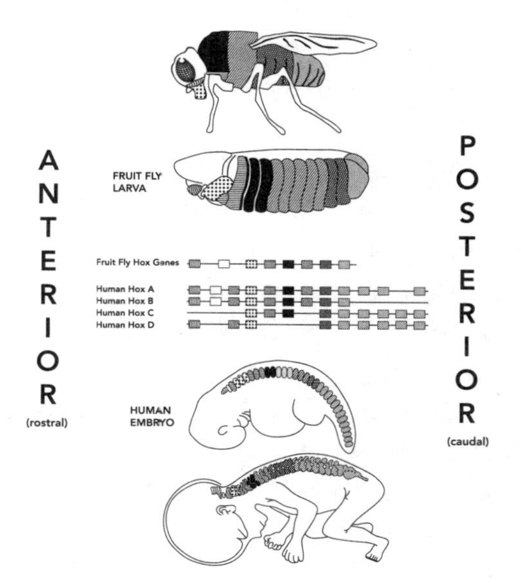

A schema showing how the same Hox genes arranged linearly on chromosomes specify corresponding body parts both in the embryos and young of humans and flies. Credit: Inna Gertsberg.

word *genetics*, had found that when things go wrong in development, they often go wrong in similar ways. The Hox genes were the original Lego kit for making all the bodies on Earth; by tweaking them as kids might do—stretching vertebrae to form giraffe necks, doubling leg segments to make centipedes, playing with different shapes of wings and mouths and tails—evolution had produced the great variety of life on the planet. Bateson had been baffled by cyclops eyes and six-fingered toes, but now it became apparent: They were simply due to quirks in Hox gene expression.

The discovery of Hox genes blew biology wide open. Nineteenth-century scientists believed that the wing of the bat, the paddle of the porpoise, the hand of the monkey, and of man, all had an underlying unity. Sir Richard Owen had given the phenomena a name, *homology*, and Darwin and Haeckel had argued that homology was evidence for evolution, not a sign of a celestial plan. But both Darwin and Haeckel had stopped at form, at what the eye could see. Now, modern evo-devo was showing that homology was deeper than that: It went back 540 million years, and all the way down to the level of genes. From jellyfish to mayflies, and from eels to humans, all living things have a common origin. When that origin settled on a number of different body plans during the Cambrian, what came next, majestically, were variations on themes.

In their home overlooking the orca waters, Jim and Lynn bring out a 1705 original of Maria Sibylla Merian's magisterial Surinam book, and we leaf through it together. There's the pineapple she drew, I say just a bit too excitedly, and the tarantula eating a hummingbird. And there, at the distance of more than three centuries, is the lanternfly, the one that scared Maria in the middle of the night, with its alligator head and misattributed cicada parent. "It's our most prized possession," Jim says, and Lynn nods her head in agreement.

When Jim and Lynn began asking evolutionary questions, the recent developments in evo-devo caught their attention. After all, to solve the question of where metamorphosis came from, they'd need to show that genes that specify form had evolved. To gain the necessary tools, they spent time with a leading evo-devo researcher at Cambridge.* Year by year, experiment by experiment, they gradually put the pieces of the puzzle together. Finally, at the age of eighty-six (Lynn) and seventy-seven (Jim), in 2022, they announced: Since control of the life cycle is conserved across species and even phyla, the three genes, *chinmo-broad-E93*, act like a Hox system—a kind of recipe for making an insect—only a temporal, rather than a spatial, one. With the genetic evidence in hand, Jim and Lynn now suggested a plausible evolutionary scenario. When early insects swam out of the seas and began to conquer land, before too long a genetic tweak in timing pulled juvenile hormone across the birth divide, expressing it *after* hatching, rather than *before*. This redirected development to a new pathway that produced insects with small wing pads, the kind that grew from molt to molt until they turned into true flying wings. And if one such creature hatched prematurely from its egg one day, it would have an advantage: With more food suddenly available, it

* Professor Michael Akam.

could grow faster. This new kind of creature looked nothing like its adult, but it grew stronger and larger, delaying the costs of becoming an adult as much as it could. After all, it takes a lot of energy to turn into an adult. And since wings and genitals only make sense if you can use them, why bother to make them before you really need them? Instead, just collapse all the necessary molts into one dramatic step—the pupa. If the advantage of becoming bigger and stronger faster outweighs the disadvantage of staving off adulthood, just like for humans, sometimes it's better to delay paying the bill.

Lynn and Jim grin. Darwin wrote in *On the Origin of Species* that "every single organic being around us may be said to be striving to the utmost to increase in numbers; that each lives by a struggle at some period of its life; that heavy destruction inevitably falls either on the young or old." And he provided an astonishing simile: "The face of Nature may be compared to a yielding surface, with ten thousand sharp wedges packed close together and driven inwards by incessant blows." It makes sense to them, Jim and Lynn tell me, that evolution would drive a wedge as strongly as possible between juvenile and adult. After all, competition between young and old would evaporate if each adapted to its own unique environment. And so, almost three hundred million years ago, growth and reproduction were safely separated, and when the flowers came, the creepy-crawly beneficiaries conquered the world. A toothy larva can live within a fruit, supremely crafted for consumption. But its adult, a procreation machine, will flit from flower to flower in search of a mate.

It's a beautiful origin story, worthy of a nineteenth-century Russian novel. But siding with the late Hinton, there are those who believe that Jim and Lynn—and Imms and Burlese and Harvey and Aristotle—are just dead wrong. To their minds, caterpillars are no kind of immature hatchlings. Like nymphs, they are creatures perfectly crafted to fit their environment. Look at their highly specialized features, they say, like eating mandibles, prolegs with hooks to hang on to branches, antennae to smell, palpi to feel, eyes to see, and spinnerets. The larva must have been born from the very last stage of the pre-adult. It's the pupa, not the larva, they insist, that's in need of an explanation. And maybe what drove the invention of the pupa is a matter of health: If a caterpillar gets rid of its gut, cleverly, it also gets rid of its gut's parasites. Another hypothesis is that it makes sense to be big and strong and full of energy before embarking on dramatic shapeshifting, since females can grow more fecund and males better at fighting as adults for mates. And so a pupa was born, making a clear separation between growth and differentiation. Related though distinct is the notion of breaking genetic correlations between different developmental stages, what the evolutionary biologist and entomologist Nancy Moran called "adaptive decoupling"; minus the modern term, Wigglesworth had already proposed this idea back in 1954. From an evo-devo perspective, selection working across the life cycle increases a creature's ability to evolve since the various stages provide more variation for evolution to play with. Whatever the true explanation, critics say Jim and Lynn are well meaning but naïve. In fact, their evolutionary scenario has things precisely backward.

I stare through the large wooden windows at the North Pacific. The community of metamorphosis experts worldwide is tiny—just a few dozen insect people around the world and a few hundred for all the other creatures combined. Why is such a small group of initiates so divided about the origin of what they study? How can they disagree over what part of metamorphosis is a novelty? Leafing through Maria Sibylla Merian's Surinam book, I suggest a possible answer: Maybe it boils down to the question of self. Is a self the shape of an animal or is it its chemistry? Is it wholeness or parts? Cultures of science can be difficult to escape from. Physiologists like Lynn and Jim follow molecules, whereas entomologists and taxonomists follow form and behavior. People who view things from above as systems often have their own unique perspective. Each group of trained scientists looks at what falls under its lamplight. But maybe a self is all of these things at once.

Jim walks over to the sink and smiles. He doesn't seem confused about any of this, though he admits it would be good to test their ideas out on more species. He insists that butterfly and moth larvae aren't immature hatchlings, as critics claim, they're "redirected embryos." It shouldn't be surprising, therefore, that they've evolved all kinds of special features to adapt to the outside world. Follow the hormones and you'll see just how it works: The job of *chinmo*, after all, is to prevent the embryo from transforming into a nymph. And so the expression of *chinmo* for longer periods after hatching would have increasingly staved off the costs of turning into adults, but must have occurred first in the early part of pre-adult development, not at its end. That's why it's the larva, not the pupa, that's in need of an explanation. To minimize the cost of the larva's rapid growth, the pupa would have just followed in tow.

Jim and Lynn remain confident that while Haeckel's "ontogeny recapitulates phylogeny" is an oversimplification, it reminds us that creatures weren't designed, but evolved. By following genes and hormones, just like their predecessor followed embryos, they know that development and evolution are intimately involved. Yes, the involvement can be tortuous; Haeckel understood that and so did Darwin. But Jim and Lynn are confident that the larvae of sea squirts and starfish are ancestral forms. So are the larvae of frogs and axolotls. There's even an example of permanent ancestral larvae—they're called fish, Jim laughs nerdily. Ascidians and amphibians are alike, but insects are different. It was insects, Jim and Lynn insist, who brought the world a truly novel invention. They alone hide their real selves until it's time for revelation. "Remember," Jim's eyes twinkle, "that *larva* is Latin for 'mask.'"

I leave the conversation dazed. It seems to me that how metamorphosis originated and why it succeeded are two different questions and that sometimes they get confused. We need to explain both the birth of the pupa and of the larva, not one or the other. That doesn't mean that they necessarily evolved together. As more time was spent growing, there remained less time to turn into an adult. And so there was a choice to be made: The creature could either add more stages to its life cycle, or cram all the stages into one, and make that one stage longer. Molting is costly, the insect physiologist Stuart Reynolds from

Bath explains to me, because you need energy to shed skin and build a new cuticle. It's dangerous, too, since insects wriggling to get out of their old armor are easier targets for hungry predators. And so the second solution makes more sense.

However metamorphosis came into this world, and whatever feature of it was novel, what is certain is that it produced a solution that was phenomenally successful: Completely metamorphosing insects include the order with the most species on Earth—the beetles, or Coleoptera. They include the order with the most biomass—the sawflies, wasps, bees, and ants—or Hymenoptera. And they also have the greatest impacts on agriculture (through the butterflies and moths, or Lepidoptera), and on humans' and animals' health (through the flies, or Diptera). Pulling apart juvenile from adult, fashioning them entirely different in form and behavior, in what they eat and where they roam, evolution turned little sea worms into striking giant peacock moths, into singing cicadas, imposter large blues, firebugs, firebrats, frantic amorous mayflies, and untold species more. It's a triumph that makes the heart skip a beat.

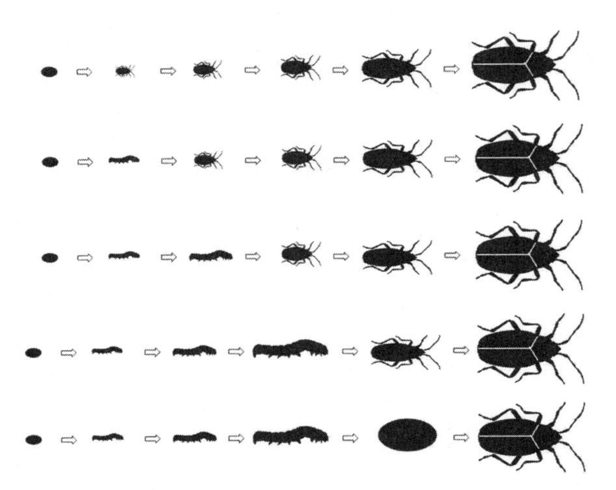

A graphic, generously provided by Stuart Reynolds, showing the evolutionary dynamic of an increased emphasis on larval growth leading to the birth of a pupa. The more time a creature has to grow, the more adaptive it is to fit all bodily changes into one concentrated stage, the pupa, rather than molt along the way.

In *The Voyage of the Beagle*, Darwin recounts how when he arrived at the port of San Fernando, Chile, the authorities had arrested a man claiming he could transform worms into butterflies. The man, a German named Renous, was accused of witchcraft. Darwin thought this preposterous. He had no knowledge of the discoveries the future would hold, and would have been tickled silly by the uncovering of the role of genes and hormones. Yet like the true father of evolution, he described the marvel better than anyone:

> There is grandeur in this view of life, with its several powers, having been originally breathed into a few forms or into one; and that, whilst this planet has gone cycling on according to the fixed law of gravity, from so simple a beginning endless forms most beautiful and most wonderful have been, and are being, evolved.

9.

THYROXINE

We're just a few weeks away from our due date, and I take my boy out to practice riding his bike in the playground. When our baby is born, I say, holding a broomstick tied to the back of his seat to steady him, it'll be a kind of metamorphosis. The placenta helped her "breathe" throughout the pregnancy, carrying oxygen and carbon dioxide between her and Mom. But the moment she finally emerges from Mommy's belly, her lungs will be filled with water. When air will flow in for the first time, she'll gasp!

Trotting behind him, I remind Abie not to look back at me but keep his eyes on the path ahead. Inside Mommy, I continue, our baby has made twice as much heat as an adult, and when she comes out she'll need to cool. Receptors on her skin will send messages to her brain that her temperature is dropping, making her body create more heat by burning baby fat. At the same time, her liver will need to spring into action. Until now it's stored sugar, and iron, but when she's born, it'll start breaking down waste . . .

Gradually it dawns on me that I'm speaking to myself. Abie is so concentrated, he's in his own world. I look at him and feel his sweetness on my skin like a burn. I want him to stay just as he is, not to change one iota. But I also want him to grow up and learn.

Remember axolotl, the salamander that refuses to grow old? We share about 90 percent of our genes with it. When the temperature in the lakes in which it lives in the Mexican highlands dips beneath a certain threshold, a cascade of releasing factors converge on its thyroid gland, making it produce thyroxine, the hormone that kicks metamorphosis into gear. But the temperature in those lakes hardly ever reaches the threshold, meaning that metamorphosis doesn't kick in. The axolotl reaches sexual maturity without becoming an adult.

And what about the toad with those chrysoberyl eyes? For him, it's not temperature but rainfall that sounds the bell; in tune with water potentials in the soil, George Orwell's friend will go ahead and morph. In other species it's sometimes light falling on the retina as the seasons change, or the presence of predators. Whatever the cue, the result is the same: Stimulated by the hypothalamus, the pituitary awakens. When the peaking thyroxine courses through the blood to different parts of the body, it seeps into cells where it flicks certain genes on or off. Tissues die en masse, others form or are reprogrammed, ushering in the drama we call metamorphosis.

The molecule thyroxine.

And so a convergence happened, in insects and amphibians,* whereby the world outside instructs the organism: The time for change has come. Cells in the brain go on to initiate hormonal cascades in the body which in turn produce a particular developmental gene pattern, and transformation begins. It's a clever device, linking organism to environment. How it was fashioned in evolution provides living things with flexibility: Turn on metamorphosis early when the ponds you inhabit are ephemeral, slow down if the pond is likely still to be there in a month; skip the larval stage entirely by metamorphosing in the egg if chances are high that predators are lurking outside. That's exactly what happens in the common coquí, a small but very loud Costa Rican frog. Most frog eggs are tiny, just fractions of a millimeter across, but the coquí's eggs are twenty times that. They have to be, filled as they are with enough yolk to allow tiny tadpoles to metamorphose. Then, as four-legged, air-breathing, tail-free adults, the coquí can hatch into the world.

So what have we learned about the molecules involved? As it turns out, ecdysone (in insects) and thyroxine (in fish and amphibians) can't attach directly to the DNA in cells, but rather to receptors on the chromosomes upstream from the actual genes involved. When these switches are bound, they initiate a precise cascade of gene activation and inactivation, and if you mutate or block those receptors in the cells of insects or amphibians, metamorphosis breaks down. Considering how far apart amphibians and fish are from insects, what scientists have discovered is really quite amazing: Even though ecdysone and thyroid hormones have different chemical structures, their receptors both belong to the very same family. They work in the same way, by repressing genes in the absence of the hormone, and activating genes in its presence. And what's even more mind-blowing is that the same is true for humans.

Evolution had caught on to something. But what is going on? The answer, once again, takes us back to the oceans. Remember the sea

---

* But also in many ray-finned fishes, like cod and flatfish and sea horses and mahi-mahi.

squirts, the ones who swallow their tails and lose their heads and break their backbones to turn into wineskin adults? Their larvae have an organ in their throat called an endostyle whose function it is to capture iodine from the surrounding ocean waters. Why iodine? Most likely because the sea squirt's forefathers in the Cambrian used it as a powerful antioxidant. But gradually, these early marine creatures discovered that iodine could also serve as a catalyst for producing organic molecules, and before long they used it to make . . . thyroxine. The highly reactive and versatile molecule was subsequently integrated into the organism's machinery of energy production and gene function, eventually becoming essential for communicating between and within cells. This gave it a role in coordinating development. Sea squirt larvae are chordates, whereas their adults are invertebrates; and thyroxine, it transpires, is crucial for turning one into the other. In fact, when scientists used a poison on sea squirt larvae known to disrupt thyroxine function, they failed altogether to metamorphose.

Why is this important? Because, Frank Ryan reminds us, thyroxine also has a crucial role in humans—two roles, actually, depending on our age. In adults it plays an everyday role controlling the rate of cellular metabolism, regulating the burning of calories, effecting weight loss and gain. It can also slow down or speed up the heart rate. But when we're embryos and when we reach puberty, thyroxine plays a different role. Binding to the very same receptors that are almost identical in structure in insects, it turns into a master development switch: Based on the pattern it creates, hundreds of genes will be turned on or off. The ensuing changes to our bodies will be dramatic, whether or not we call it human metamorphosis.

In adolescents and young adults, for example, thyroid dysfunction may lead to infertility. It can disrupt the menstrual cycle in females, and—if they get pregnant—lead to premature, even stillborn, births. It is as if, on the final molt, an insect's hormonal regime malfunctioned, preventing it from reaching sexual maturity. But more dramatic than any of this is the influence of thyroid hormones on fetal development. Already in the 1980s it was known that too little thyroid hormone in a human fetus can lead to severe mental retardation. Later, genes were found in the fetus brain that are affected by thyroxine, two of which play crucial roles in mammalian development. From mice to humans, thyroid hormones cross the placenta into the brain of the fetus developing in the womb. During normal development in mammalian brains, cells "decide" whether they will be functional nerve cells (neurons) or connective (glia) cells before migrating to the cerebral cortex. And what scientists found was that at these crucial stages before migration, the amount of thyroid hormone within the cell determines which.

I think of Abie again. The normal development of the organ that allows kids to grow to become scientists and bikers and parents, the very organ that makes us feeling, thinking, moral beings, is crucially shaped by thyroid hormones. In fact, there's a simple system of just two enzymes that control the levels of thyroid hormones within the cell, as opposed to the level coursing through the fetal body, in order to get things just right: One enzyme degrades the hormone, while the other activates it. The same system exists in frogs. No wonder the man who figured out how it works in mammals started off taking orders from a fish and wildlife manager before moving on to study human brains.*

The thyroxine thread suggests a probable scenario. Around 370 million years ago, the vertebrate line produced a new branch, the amphibians. Having to live both in water and on land, this branch evolved a new strategy to best adapt to the challenge—a strategy we call amphibian metamorphosis. The vertebrates who became reptiles and mammals, on the other hand, didn't need to go to such lengths. Instead, they abandoned metamorphosis, but put to work the very same existing ingredients—hormones, receptors, genes, and nerve cells—to safeguard their normal development. As the mammals evolved, they eventually produced humans. And, relative to frogs who lay eggs, the fetuses in human mothers' wombs had more time to develop more complex head parts. The schism seems dramatic—

* R. Thomas Zoeller, emeritus professor of biology at the University of Massachusetts in Amherst.

metamorphosing creatures versus non-metamorphosing ones—and would go on to produce an Ovid in our lineage, a Goethe and a Strauss. But in reality it's nature playing with its inherited biological hardware. This tinkering is what helps us all survive.

People think sea squirts are boring, but their endostyle is a good candidate for the evolutionary forerunner of our thyroid. Perhaps before the forerunners of sea squirts could produce thyroxine, they didn't metamorphose. Williamson, the planktologist from Liverpool who hit his head on a rock, believed that an adult marine chordate that looked like a tadpole hybridized with an ancestral ascidian, thereby producing the sea squirt's unique life cycle. But maybe it's the other way around: Chordates originated from primitive ascidians whose tadpole larvae didn't metamorphose, perhaps by simply failing to respond to cues from the environment. That way, larval features would have been carried into adulthood, together with the already integrated thyroxine. Gradually, in parallel with the evolution of the primitive brain in marine animals, a novel "thyroidal" structure appeared as an adaptation that ultimately enabled the transition from the iodine-rich ocean to the iodine-deficient terrestrial environments that beckoned life beneath the waves. In fact, just like a sea squirt, a strange-looking jawless fish with suckers for a mouth called a lamprey also has an endostyle, at least for the three to seven years it exists as a larvae. Over a period of four months, the lamprey's endostyle slowly turns into a fully functioning thyroid gland, which brings about metamorphosis. Lampreys are thought to be living representatives of one of the most primitive vertebrates. And so they too serve as a link between humans and the sea.

It might have all begun, 540 million years ago, with a Cambrian sea squirt–like creature snaring a bunch of antioxidant iodine. Gradually, around the time the amphibians appeared, a creature evolved into a lamprey-like vertebrate, in the far reaches of a Devonian sea. It was then that a tetrapod climbed onto land, looking rather oblivious. Buzzing around it were insects that had already learned some of its ancestor's metamorphic tricks: They'd lost the thyroid system, but had gained the analogous ecdysone system (amazingly, scientists have shown that thyroid hormones can act like juvenile hormones to induce metamorphosis in insects, a true case of convergent evolution). Over time, the tetrapod evolved, from a land-dwelling crawler, to a monkey swinging in the trees. And before it knew it, it was embryonic you and adolescent me.

It's tough to know what really happened, and we may never know for sure. Trying to figure out where we came from a century and a half ago, Charles Darwin offered a tantalizing possibility. After all, he was well aware that despite the differences, the tadpoles of frogs looked a heck of a lot like the tadpoles of sea squirts. Perhaps one turned upward, more than just metaphorically, while the other turned downward. In his *The Descent of Man* he wrote: "At an extremely remote period a group of animals existed resembling in many respects the larvae of our present ascidians, which diverged into two great branches—the one retrograding in development and producing the present class of ascidians, the other rising to the crown and summit of the animal kingdom by giving birth to the vertebrata." Darwin, of course, knew nothing of thyroxine. There is grandeur in this view of life.

My boy looks at me from his bike and shrugs. Soon he and I will be outnumbered, two to three, that's what seems to have stuck from our talk. Abie has a point. But more than family gender politics, all I can think about is that during the last nine months, our baby's brain has been growing, adding 250,000 neurons to its mass every second. When she arrives her brain will have 100 billion neurons, as many stars as in the Milky Way. But the size of her brain will be just a quarter of that of an adult. One way it will grow is by creating more connections: At birth the number of synapses per neuron is about 2,500, but by age two it'll be closer to 15,000. All the while, her axons will be myelinating, becoming more efficient at sending information. Her cerebral cortex will balloon, and her brain will become better at mediating between her inner self and the environment. What a miracle! By expelling them from their mothers' wombs somewhat early because of our narrow pelvis, evolution has given us touchingly vulnerable babies. But in its wisdom it has also designed them to be able to learn from the world.

Taking advantage of my daydreams, Abie dashes home on his bike. I run after him, wildly, but am surprised to see his wobbly wheels settle into a confident furrow in the ground. Tears start running like rivers from my eyes. Searching for the feeling of what it's like to be a four-year-old, I come up short. My memories are like islands in a sea of forgetfulness: the smell of my grandmother's chicken soup, sun rays falling through my bedroom window, the texture of a red cape I once wore to be Superman. Imperceptibly, the threads that tie us to who we once were disintegrate, and the past sails away into a white mist. I don't know how or when it happened.

Reaching the gate ahead of me, Abie confidently throws his bike on a bush, and runs through the door laughing. I follow him inside. Yaeli is sleeping on the couch, her arms and legs rolled up around her huge tummy like afterthoughts. Is she the same person I fell in love with and married? Trying to escape my thoughts, I lunge toward Abie, but he squeaks beyond my reach. You won't be the baby soon, I say, an avuncular forewarning. At a safe distance away from me, he puts his finger to his mouth and giggles. His eyes smile as he whispers:

SHHHHHH.

10. ROSETTA STONE

On occasion, Jim Truman admits, he finds himself thinking about identity. After all, in a fly, almost all the cells of the maggot die, and the adult is formed *de novo*. One might almost say that the insect is actually two individuals separated in time.

Except that the brain is special. During metamorphosis in flies, the brain is spared from complete turnover. It isn't crushed, or shriveled up, or wholly replaced—it's mainly just rewired. And since the brain is the seat of identity, its persistence from one stage of the life cycle to the other puts things in a different light. At least that's what Jim believed—about humans, and also insects. Others had shown that a tobacco hornworm caterpillar exposed to a particular odor at the same time as it was given a tiny electric shock, learned to stay away from that odor. More dramatically—something Pavlov could never have checked on his dogs—once the caterpillar spun a pupa and finally hatched, the adult moth, too, stayed away from the smell. This was incredible, but there was a caveat: If the caterpillar was exposed earlier rather than later in its life to the odor, when it became an adult the aversion disappeared. Now the unpleasant memory was wiped out, as though it had never existed. It seemed obvious, therefore, that recall following metamorphosis involves regions of the brain that aren't produced until late in larval development, but what was going on? Looking more closely than anyone before him at the brain of a fly, Jim hoped to find the answer. Maybe he could detect anatomical, not just behavioral, evidence that memory traces could survive metamorphosis. Maybe there was a hard biological truth behind this supposed magic, facts that could explain how lessons learned by the fly when it was a maggot might serve it when it transformed into an adult, even if it really did look to the naked human eye like two completely separate individuals separated in time.

In retrospect, Jim's fascination had been an obsession. The brain collection of frogs, lampreys, and hamsters in grade school a prelude to his fascination with the flighty virgin mosquito turned into a calculating mama in Notre Dame, itself preparation for his work on biological clocks turned on and off by hormones at the old missile silos at the field station. Everywhere Jim looked, he saw brains and behaviors around him; for whatever reason, it was simply the glasses he wore to see the world. On one occasion, he and Lynn decided to breed their quarter horse mare, and Jim tried to gauge where she was in her cycle. When he thought the time was right, he had a friend bring over her Arabian stallion. But the mare became agitated, and violently rebuffed the stallion. A few more weeks of close monitoring finally brought the desired result; at the stallion's first whinny the mare became as calm as a lamb. Gone was the hostility, gone were the antics. Jim knew: All the conditions had been precisely identical on both occasions. The only difference was that more estrogen was now coursing in her brain.

We often think that brains are in charge, clean command centers sending instructions to the body, detached from all the corporeal mess. In modern times historians have argued that this may have something to do with how, ever since the 1950s, computers have become such powerful metaphors. But brains live in bodies, and need to be aware of the body's requirements—whether it's more estrogen to get a mare to mate or more ecdysone to make an insect molt. However much we view them as autonomous, brains evolved to facilitate survival, not the other way around.

And so, during the early days at Harvard, Jim started asking questions. Sure, adults have different goals than juveniles, so need different brains to help serve their bodies. But how do hormones turn a larval nervous system into the nervous system of an adult? Working on the tobacco hornworm *Manduca*, he'd seen that sometimes hormones had a direct and almost immediate effect: Inject a caterpillar with eclosion hormone, and within minutes it would shed its cuticle and start to molt. This was called a "releaser" effect of a hormone, and Jim would spend years looking for the neural targets in the brain to figure out how it turned on the complex molting behavior in the moth. But there was another way in which hormones could affect behavior: by slowly resculpting the nervous system so that it could carry out new tasks in the future. An example of this "primer"

Jim and a friend, 1974.
Courtesy of Jim Truman.

effect was the way the hormone ecdysone first prunes the neurons of an insect, then regrows them into their adult form, in two separate waves during metamorphosis. Clearly, the nuts and bolts of how this worked in moths and humans were different: Whereas insects use ecdysone and juvenile hormone, humans use estrogen and testosterone (and corticosteroids), and our brains are much, much more complex. But the emergence of "evo-devo" made it easier to see that these two systems shared common underlying principles. How does estrogen turn a hostile mare into a receptive lamb, or testosterone a boy into a man? With moths, Jim had built the only invertebrate system studying steroid action on the brain, and he wanted to find out.

Here was the triumvirate: neural death, neural remodeling, and neurogenesis (the process by which new neurons are produced). What kind of convergence of these three roads turned a juvenile brain into an adult one? The best way to find out was to track neurons in the caterpillar and see where and in what shape they ended

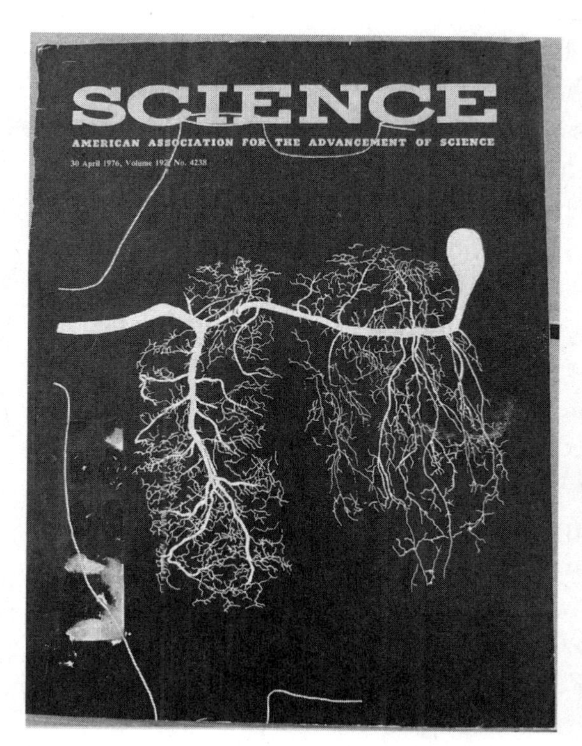

The original 1976 *Science* cover with glued subscription address label markings in lower left, showing a motoneuron of *Manduca* reorganizing during metamorphosis. "It's been hanging on our wall for almost fifty years," Jim says, "an historical document!" Courtesy of Jim Truman.

up in the moth. But these were the seventies, and it would take many more years until researchers could do this in the same animal. It was possible, though, to stain a small set of motor neuron cells by sending dye up the cut stump of nerve branches that contained their axons. That way, though not tracking the selfsame neuron through metamorphosis, one could take "snap shots" in a series of successive caterpillars and moths. By sucking nerves into a glass pipette, and later using more reliable technology, slowly but surely Jim was figuring out how a brain transforms.

"We fell in love with Africa," Jim says, as we walk down to see harbor seals in the bay near their home. Lynn is well behind us, walking slowly with ski poles, and once in a while Jim cries out to see how she's doing. "Every year we'd spend three weeks there," he says, "unwinding from the crush of science. It began in the eighties, sitting outside our tent in the Maasai Mara, looking at a pride of lions and thinking that our ancestors may have been at that very spot and witnessed the very same scene." It was a feeling of coming home, and also of deliverance: The hand of man could disappear from these plains entirely, and the animals would fare just fine. But Africa was also Egypt, where a great human civilization once flourished. To reclaim its lost culture, nineteenth-century scholars had used a cipher, the kind that allowed hieroglyphs to be read by comparing them to the known languages demotic and Greek. The brain, Jim thought to himself, was a much older mystery. Maybe the fly could serve as its Rosetta stone.

"Retirement" seemed his best shot at the dream: When Jim and Lynn hit the required age in 2007, they moved to the Howard Hughes Medical Institute in Virginia. Janelia, as the research campus was called, was making major investments at the time in *Drosophila*, with the goal of a complete wiring diagram, or "connectome," of its nervous system, along with the creation of genetic tools that would allow manipulation of individual neurons in the circuitry. Jim, like Lynn, had already learned that *Manduca* would not be up to the task. Now, with his long-standing obsession with how nervous systems diversify and evolve, Jim couldn't have dreamed of a more perfect fit. Saddled with almost unlimited resources, and time, Janelia and its flies would be an opportunity to finally pull back and take a look at the big picture.

Jim knew well: There are many more species of insects than human languages, but all have a similar nervous system. This is because they all have similar body plans—a head, a thorax, and an abdomen. Along this axis, starting with the brain, lay the ventral nerve cord, the insect equivalent of our spinal cord, and each part played its role: The brain helps insects see, smell, remember food locations, communicate, navigate, mate, maybe even feel. Just below it, a clump of nerves, called the subesophageal ganglion, controls the antennae and neck movements, as well as the mandibles that help insects eat. Below that three independent nerve clumps, called thoracic ganglia, control locomotion (three pairs of legs and a pair of wings), and below them abdominal ganglia control respiration, circulation, digestion, and everything having to do with sex. Insects are manifold; there's an enormous variety in the number of abdominal ganglion, and an enormous variety in the spacing of thoracic ganglia (in some insects they're completely fused). But at the level of basic nervous system architecture, all insects are basically clones.

Drosophila could be a cipher, then, a kind of blueprint for how all insects build nervous systems. It could teach us what it takes to

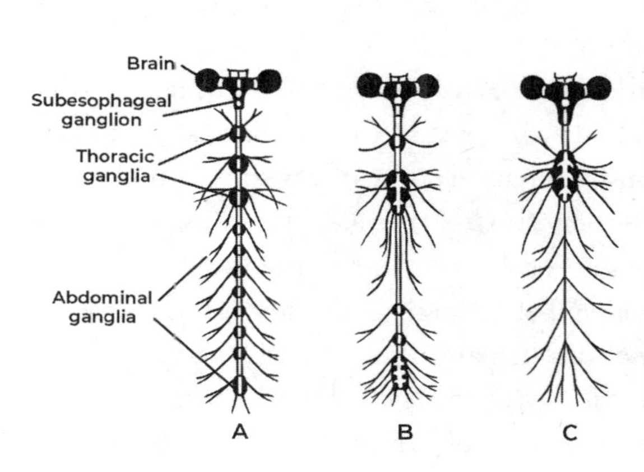

Despite the large variety of insects in nature, the basic plan of the nervous system of all insects has four parts: a brain, a subesophageal ganglion at the stem of the brain, three thoracic ganglia, and abdominal ganglia. *A*, *B*, and *C* represent three different body plans illustrating this unity. Credit: Inna Gertsberg.

create a new function, like flying, when evolution comes knocking at the door. The three thoracic ganglia that controlled a fly's locomotion seemed like a good bet to start with. Compared to the brain they were relatively simple, and each ganglion was discrete, like a bead on a necklace specially evolved to meet the requirements of its segment. With these advantages in hand, Jim could ask what kind of architecture and computing power a wriggling maggot needs to transform into a six-legged winged fly. He knew that every animal needs different components that allow it to react to its environment: motor neurons to carry information from the brain to the body, sensory neurons that went the other way, interneurons that made connections between the two. To figure out just how such a system was built, you'd need to figure out how it developed. It was long known that cells called neuroblasts, just like stem cells, provided the initial stock from which a nervous system was grown. The big question was: Would knowing how neuroblasts code for the "language" of becoming a fly also help you figure out how a locust becomes a locust, or a butterfly a butterfly?

Jim believed the answer was yes. One major reason for his confidence was that ever since the mid-1980s it was known that neuroblasts are ancient, and have remained pretty much the same for almost half a billion years. That meant that a fly and locust and butterfly, but also an earwig or firebrat, were all made from the same components. Another reason was that the process of building an insect nervous system was modular. Others had shown in the locust that each thoracic ganglion came from only thirty-one initial neuroblasts, thirty of which are paired and one unpaired. What was striking was that the rules of development had been fixed all the way back at the base of insect history: Each neuroblast had to divide repeatedly to make a set of precursor cells, which then divided themselves, just once, to produce either an *A* or *B* daughter cell, each with a different fate. But the precursor cells don't divide all at once; instead, throughout development, some divide early and some late. That way, the thirty-one neuroblasts each end up producing two distinct "hemilineages," and the timing of the division of daughter cells from the stock of precursors in each lineage, and the number of cells produced, determines the rate and shape of development.

Working on flies at Janelia, it soon became clear to Jim and his team that the hemilineage modules are not just developmental but also functional; in other words, they didn't just determine how the fly grew, but also how it behaved. Some build flight circuits, others build leg circuits, still others leg-wing circuits that coordinate the two, allowing a fly to land and take off. Using the rich *Drosophila* tools available, Jim discovered that he could adjust the number of neurons needed for a particular function "simply" by changing the number of divisions made by the appropriate parent neuroblast. This was huge. A fly—and a locust and a butterfly and cockroach and ant and ladybug—are all built differently for their unique and different lives. But if the parts and rules by which they're built are conserved in evolution, figuring out how it works in the fly would provide a cipher. Suddenly Jim could ask questions like: What set of neuron types are necessary for making a creature that can coordinate between its leg movements? Or, how much does a neuroblast need to dial up the number of neurons it produces to gain enough computational power to allow, say, a fly to make more refined spatial maps of its world? He could also ask evolutionary questions: How did the nervous system change to support the evolution of flight, for example, or what did the flight modules do before flight was invented? By creating the first full library of the anatomy and function of every module in the ventral nerve cord, he would provide a toolbox of neural types necessary to make a thoracic ganglion. At long last, Jim had his own Rosetta stone.

These are early days in deciphering the development and evolution of insect nervous systems, and many of the questions Jim is asking now don't yet have answers. A fly has 5,000 neurons in each of its thoracic ganglions and a million in its brain, a far cry from the 86 billion neurons we humans have. Even still, it'll take years before we know how a fly thinks, if we ever do. The brain is one thing, but the mind is quite another.

In what aficionados have called "The Paper," Jim has recently taken the closest-ever look at what happens to the brain of a fly during metamorphosis. It took fifty years, but using genetically engineered flies whose cells can be cleverly marked with green and red fluorescents and viewed under the microscope,* he was finally able to trace individual neurons through the process. The question was: How does a brain whose only concern is food turn into a brain obsessed with mating? And what Jim found is astonishing: While almost all of the cells making up the body of a fly die during metamorphosis, the brain cells themselves don't die in large part, but are rewired. Brand new legs and wings and genitals of an adult fly emerge from set-aside imaginal discs, like perfect sculptures rising from a desert of dead cells. But the brain stays almost exactly the same, only significantly remodeled. It may seem as though an insect just melts inside its pupa, since pricking it results in a goo oozing out, as both Swammerdam in the seventeenth century and Fabre in the nineteenth, experienced. But the truth is that it's a delicate assemblage. Recently, 3D Micro-Computed Tomography (CT) images scanned in a Nikon Metrology HMX ST 225 system have shown in high resolution a pupated maggot, step-by-tiny-step, turning into a fly. Fabre and Swammerdam (not to mention Haeckel, Maria Sibylla Merian, and Aristotle) would have been dumbfounded. Inside—especially the neurons that wire the brain—everything has a fixed position, like parts in a mechanical clock. Prick it and it will pop.

To see how this works, Jim looked at the mushroom bodies, a part of the brain that plays a crucial role in learning and memory

---

* Green fluorescent protein, or GFP, is a protein that turns green when exposed to a certain wavelength of light. A number of groups in the 1990s figured out how it could be inserted into a cell to track how that cell behaves and develops; among them, a group at the University of Washington figured out the protein's structure, the silver statue of which Jim pointed out to me on the grounds at the Friday Harbor Labs.

(especially related to smell). He knew that how a fly makes a nervous system is a process older than metamorphosis, which meant that once neurons were made for the maggot, they couldn't be changed anymore for the adult. Instead, things would be reshuffled. Mushroom bodies are clumps of neurons that shoot long axons in parallel lines, like guitar strings, to communicate with other parts of the brain. In a maggot there are ten "frets" along these strings, each comprising a distinctive computational compartment that allows a fly to associate a certain smell, say, with danger, or perhaps opportunity. Input and output neurons weave through these strings to strengthen these memories; if the maggot comes across that smell a second time, it knows whether to move toward it or recoil.

What Jim found was that during metamorphosis, seven of the ten "fret" compartments survive into adulthood, whereas three cut their ties, shed their old identities, and migrate to other parts of the brain. This was in keeping with Jim and Lynn's theory that the larva is a new development in evolution—the computational compartments had been "on loan" for a while to the simpler maggot, helping it in its struggles, and were now returning to their original roles in the ancestral adult brain. But what was most incredible was that all seven "frets" that did become incorporated in the adult, *as is* from the maggot, completely unplugged their input and output neurons first. For a brief while, less than a day, the fly existed in a twilight zone between juvenile and an adult, disconnected from its past and all experiences, possessing no self. To mature, every single fly in the world does what Buddhist priests spend a lifetime trying to accomplish. Then suddenly they are rewired again, ushering a new being into the world.

Here's the catch: Jim found no anatomical evidence whatsoever that memories made in the maggot can make it into the adult brain. Whatever happened in the fly's previous life was wiped clean from the slate. Other scientists had shown that in other insects that isn't always true: The butterfly that stays away from a toxin to which it was exposed as a larva is one example, mentioned before. Another are the many insects that hatch on a particular plant, say an apple tree, and somehow remember to return to apple trees when laying their eggs, even though they have nothing to do with them as adults. Why do some insects retain some memories, whereas flies seem to be born, literally, from a state undone?

One possibility is that associative memories really don't survive metamorphosis, but other kinds of memories, kept in other parts of the brain, do. Jim has another theory: It's simply a numbers game. After all, the identity of a neuron depends on its lineage, on which of the initial neuroblasts produced it, and on the relative time of its birth. In insects that undergo an incomplete metamorphosis—and whose nymphs therefore look like miniature adults—all of the neurons that go into making a mushroom body are already produced in the embryo, so the hatchlings are fine. But in a completely metamorphosing creature like a butterfly, the evolutionary trend has been to bring about hatching as soon as possible. That means simplifying the larva, including the development of its brain. Flies stand at the extreme end of this trend: It takes a fly egg just twenty-four hours to produce a maggot, but the neuroblasts in that hatchling will have only made between 10 and 15 percent of the neurons needed to build an adult brain. Butterflies, on the other hand, need longer to make a caterpillar—between four days to three weeks or even more, depending on the species. No wonder flies don't carry memories into adulthood, whereas butterflies seem to be able to, at least to some extent.* The more complex the larva brain is, the less it needs remodeling and rewiring, and the greater the chance that memories will survive.

Using the molecular trinity, Jim hopes to "hack" the fly brain one day, to genetically manipulate neuroblasts to make an adult brain cell in a maggot. Maybe that way he'll produce a fly that remembers its past life; certainly it will help us understand how the very same cells change their identity to make a flying sex machine from a wriggly glutton. But Jim's theory of a numbers game makes you wonder: Is identity the same as memory? If it's more than that, what besides

* A dangerous toxin is dramatic, something to remember. Whether other kinds of memories survive we do not know.

memory is identity really made of? And what turns identity into a sense of self?

Scientists studying the brilliant-thighed tropical poison frog from South America have shown that it retains personality traits across metamorphosis: Despite the dramatic restructuring of their brains, despite the habitat transition and completely new anatomy, bold tadpoles turn into bold adults and timid tadpoles remain timid grown-ups. According to Jim and Lynn, this shouldn't surprise us: Amphibian tadpoles have ancestral brains, unlike insects' larvae, whose brains were simplified when the larva was invented late in evolution. And so it stands to reason that a shy tadpole will turn into a shy frog, just as a kid moving away from the world, or against it, will continue to do so when he or she matures.

I think of Shaizee and her love for pickles. She doesn't believe me when I tell her that she used to hate them. No way, she says. No chance. Impossible. Forget it. Gerit Linneweber, a neuroscientist and developmental biologist, laughs. In humans, brains are almost fully grown by the time we're five, but an enormous amount of pruning of axons and rewiring continues to go on well past adolescence. According to Gerit, Shaizee the pickle lover is really an entirely different person from Shaizee the pickle hater—no wonder she can't believe it! But I wonder. From the moment she was delivered, she's had the same naughty twinkle in her eye, that same blend of humor and poise.

"Self" is way too fuzzy a concept, Jim says. Instead, it's better to concentrate on what can be measured. Short of solving any existential puzzles, figuring out fly brain development in exquisite detail will help unlock problems mysterious enough: how jellyfish and starfish, cicadas and mayflies, eels and axolotls become adults. Maybe it'll even be useful for learning how a pickle-hating baby can turn into a Cleopatra or Queen Elizabeth (both were piccalilli crazy). That's down to cell identity, a scientific concept. The "self," on the other hand, is in the eye of the beholder.

It's getting late and I have a ferry to catch back to the mainland. "Do you think you've changed over a lifetime spent studying metamorphosis?" I ask for the third time. "Not really," comes Jim's reply, just like his previous ones. He and Lynn started out as kids who loved nature, he says, and the only thing that's changed is that they've gained tools and experience. "We're not particularly social, we're just very curious." That's why, over nearly seventy years, both Jim and Lynn have never left the bench, a rarity among scientists who tend to graduate into managers. "We didn't have children," they smile, "so the high point of our week has always been Saturdays. That's when we can go to the lab quietly, with no one around, to play with the animals we love."

We put on our coats and open the front door, laughing. To the left of us, the Pacific Northwest is like a glass mirror of the evening sky. Jim helps Lynn into the car, a slow, rehearsed dance of love, then flits around to the driver's seat. Sitting somewhat crunched in the back seat of the small Chevrolet hatchback behind this lovely octogenarian couple, I think about how the history of science is often portrayed as one great discovery after another, as an idealized march of progress. How funny that the two greatest-ever experts on metamorphosis in the history of humankind have returned full circle to Aristotle.

I allow myself to doubt that Jim and Lynn haven't changed, as they avow. Yes, their lives are a testament to the fact that humans don't "unfurl" biologically, like an acorn turning into an oak tree or a caterpillar into a mating butterfly. We are free to make our choices in life, and we exercise our potentials in many and wondrous ways. But all living things change, and Jim and Lynn are no exception. They insist that speaking of humans undergoing metamorphosis can only be a metaphor; as much as people transform during their lives, you can recognize them as themselves, which isn't true of a caterpillar that becomes a butterfly. Maybe they're right, or maybe humans change most from the inside. In any case, like Maria Sibylla Merian and Ernst Haeckel, Lynn and Jim are faithful signs of their times. Kant put it best: "Natural science will never discover to us the inside of things, i.e. that which is not appearances," he wrote. "But natural science does not need this for its physical explanations." It's what matter *does*—not how it feels or why it was created—that modern science promises to reveal.

Lynn and Jim have both spent their lives with insects, but each had their own quest. She wanted to decipher the life cycle, whereas he wanted to unlock brains and behavior. When Jim was her student, they coauthored papers, but over time their interests diverged. Jim calculated that only 14 percent of his, and 15 percent of Lynn's, hundreds of papers have been written together. Something *has* changed, he writes to me later, in what feels like an intimate confession. We've become our closest collaborators. Now that we've grown old, everything we do, we do together.

I step out of the car to the sound of the ferry horn, and we give each other hugs. Walking down the wharf, I turn back for a fleeting moment. Jim and Lynn are standing there, waving goodbye.

Jim and Lynn and the author on the balcony of their Friday Harbor island home, February 2024. Photo: Oren Harman.

11.

SOL

Our daughter Sol was born at dawn, just as the sun rose. Yaeli gasped when her tiny head popped out, and I could see two blueish pulsating strips—the umbilical cord wrapped twice around her neck. In the split of a second, the midwife performed what seemed like pure acrobatic magic: a shoulder pushed, an ear squeezed, a hand raised and . . . PLOOP. Suddenly, she was there, wrinkled and crying. The pictures would later show that she was purple, coated with cheesy-white vernix. But to me she was a majestic large blue emerging from a chrysalis.

Observing the same marvel, humans have explained metamorphosis using three different orders—Philosophy, God, and Science—describing it variably as proof of human perfection in a causal universe, a reflection of divine will, a molecular symphony, a road taken on a blind evolutionary path. Looking at the natural world, metamorphosis does seem so strange a phenomenon, at once nearly universal and idiosyncratic. But the imprecision of definitions of metamorphosis throughout the ages, the degree to which the meaning of change is always in the eye of the beholder, make it clear just how intimately science and culture walk hand in hand. When all is said and done, how we define metamorphosis is tied to our ever-changing views of our human selves. Which makes following a blue butterfly from one's childhood seem a little less crazy.

The search for self is a journey we all recognize. We ask: Are we the same person? The child and adolescent and young adult we once were? Two millennia ago, the Greek priest and biographer Plutarch asked the same question, speaking of the mythical ship, the Argo, that bore the leader of the Athenians on his adventures and brought him safely home. Over the centuries, the Athenians took pains to preserve the Argo, replacing the old planks with new timber whenever they decayed. But once every plank had been long since replaced, was the Argo still the Argo?

Today, we know that our body's cells die at the rate of tens of billions daily; by the time we're old, few of the actual cells in our body are the ones we were born with. And yet most of the time we feel no doubt about our identity. In the seventeenth century, John Locke asked his readers to imagine the souls of a prince and cobbler being swapped, so that all the memories of one now became the memories of the other. Surely, Locke wrote, it isn't the body, but the memories, that make a man, and the cobbler would now really be the prince, and the prince the cobbler. Many of us agree, speaking of "souls" or of a "true self." Some of us lament losing our true self at a young age when we start seeking to fulfill the expectations of others, or otherwise fight them, allowing a pseudo-self to take root.

In our own lifetime, a don at All Souls College in Oxford took Locke's thought experiment one step further. Very much unaware of *Star Trek*, Derek Parfit imagined a teletransporter that could beam us up to Mars. Press a button, lose consciousness for what seems like a second, and awake on the red planet. What's really happened is that a scanner on Earth has destroyed your brain and body, recorded the exact states of all your cells, and transmitted the information into space at the speed of light. Three minutes later, a replicator on Mars will create a brain and body identical to yours out of new matter, and it'll be this body into which you wake up (don't worry, you'll be beamed back to Earth if you want to).

But wait.

Now Parfit imagines a scenario in which you press the button on Earth and nothing happens. Leaving the transporter somewhat miffed, you come across a technician who explains that there's a new scanner now that records your cells *without* destroying your brain and body, and that if you stick around a bit, you can talk to yourself on Mars. As you're wondering how you can be on both Earth and Mars at the same time, a white-coated man gently ushers you into his office. "I'm afraid that we're having problems with our new scanner," he explains, eventually telling you that while it's copied and transmitted the state of your cells perfectly to another body on Mars, it unfortunately injured your heart in the process, and you should expect to die in the next few days.

Parfit was trying to say that we are nothing more than the state of the cells of our brains and bodies; that the interrelated series of mental and physical events that they produce make you you and me me. There is no entity that is you or me above and beyond this. And so what we call "identity" (or "soul" or "true self") isn't really what matters for survival. Instead, whether I continue to exist or not depends on whether some future person will be psychologically connected to my present self. And this is a question of degree, not of all-or-nothing. We hold memories of our past and desires for our future, but these memories and desires change over time. When we die, the person with our name ceases to exist, but there will be other people who have memories and desires and experiences and thoughts—who remember us. Friends and loved ones and even strangers who were influenced, or touched, by us. Only when their memory of us, or the memory of us that they transmit to others, is no longer, do we cease to be (though some of us may be discovered again in the future).

It may sound intuitive, but Parfit was making a dramatic claim: that the boundary between me today and me tomorrow, and between me and others, is more fluid than is usually implied by Western notions of "the self."

By showing that humans are constituted by nothing more and nothing less than their changing brains, bodies, and psychologies over time, modern science has done away with the unitary "self" as it has long existed in the Judeo-Christian tradition. The atheist philosopher Daniel Dennett has called the self a "convenient fiction," comparing it to the center of a hoop, a point of nothingness in the air that nevertheless provides a core of gravity. Others speak of a permanent "I," a result of a narrative we tell ourselves, juxtaposed to a momentary "me" which is the thing that experiences. Or maybe the self is a kind of commonwealth, as Parfit suggests, following in the footsteps of David Hume—a bundle of traits, including memory, but also hopes and desires, all of which are forever in flux. Maybe we can travel to and away from that commonwealth, viewing it from the outside as well as from within. Maybe we can elect new leaders every so often, or change the rules of engagement a bit. Or maybe, if we're really brave, we can finally learn to dissolve the self completely, as the Buddhists and the flies teach: Only then, when we accept that there is impermanence in life, can we ever truly exist.

A few years ago *The New Yorker* ran a piece titled "Are You the Same Person You Used to Be?" With jest mixed with a bit of angst, the author asked: "Are we the same people at four that we will be at twenty-four, forty-four, or seventy-four. . . . Is the fix already in?"

In the seventies, a group of psychologists in New Zealand launched a study to look into that very question, observing and interviewing kids (and their family and friends) at three, then again at five, seven, nine, eleven, thirteen, fifteen, eighteen, twenty-one, twenty-six, thirty-two, thirty-eight, and forty-five. Fifty years on, the Dunedin Study—named for the city in which it took place—provides an interesting answer. Having placed three-year-olds in five separate categories, the study found that at eighteen years of age, patterns begin to emerge. The "confident," "reserved," and "well-adjusted" kids blurred into one another, and they were the majority. But the one-in-ten initially scored as "inhibited" and the one-in-ten initially scored as "under-controlled" stayed truer to themselves. People in these two smaller categories were further described as "moving away from the world" and "moving against the world." The first group was more likely to live low-key, guarded lives, whereas members of the second group were more likely to be fired, suffer from substance abuse, or take up gambling. And these two dispositions endured.

There is a sense in which we curate our lives, and a lot of this has to do with how we react to those who surround us. Someone with a disposition to move away from the world will more likely interpret friendliness as overfriendliness, and someone moving against the world might read shyness as apathy, or disdain. Each type of personality will begin to build a niche for itself, favoring social situations that reinforce its disposition. None of this is written in stone, of course: Meaningful relationships, in particular, have a way of altering one's disposition, building new bridges to the world. But in large part, character is a self-fulling prophecy, which makes it difficult to change.

Yet we are more than just our dispositions, and what you are like is not the same as who you are. In an important way, what you do in life, not how you feel, defines you. No less meaningful is your own self-understanding. Some people, as the philosopher Galen Strawson argues, need to tell a story about their lives, whereas others experience life as episodic; the first have been called "continuers" and the latter "dividers." Strawson testifies that he's closer to the episodic end of the spectrum, having no sense of his life as a narrative with any form, and taking no interest in his own past. (He is akin to the narrator of a recent Richard Ford novel who says: "Everything's, of course, narrative these days. And I didn't particularly give a shit about mine.") But Strawson is famous for arguing that free will is a fiction, and his father—himself a well-known philosopher—was famous for defending the opposite. A perceptive biographer would therefore notice that

Strawson fits on a longer arc with a discernable plot. People can feel discontinuous on the inside but seem continuous from the outside, and the same holds true in the reverse.

Humans are story-telling creatures, even when their story is that they have no story to tell. And a person can be inconsistent in a very consistent way, if being inconsistent gives them a sense of value. After all, telling yourself that you are changeable is telling yourself that you are free, unencumbered, unpredictable, not set in your ways. It may also mean embracing the persona of someone vulnerable, who makes choices and transforms themself—and that self-image may well help to guide you. The same is true for the person insisting that they've always been themselves all along; for that person, being true to oneself might be either hard or easy, and for either reason, desirable. Choosing to be either a "continuer" or "divider" is a disposition. But it might also be an ideological choice.

Rumi, for his part, found solace in metamorphosis:

I died as a mineral and became a plant,
I died as plant and became animal,
I died as an animal and I was human.
Why should I fear? When was I less by dying?
Yet once more I shall die, to soar
With angels blest; but even from angelhood
I must pass on: all except God perishes.
Only when I have given up my angel-soul,
Shall I become what no mind has ever conceived.
Oh, let me not exist! for Non-existence
Proclaims in organ tones, To God we shall return.

Not everyone has a God, as Rumi and Maria Sibylla Merian did. Nor does everyone believe, as Haeckel and many of the Romantic

Movement thought, that spirit and matter are one. If science teaches us anything, it's that the certainties we hold are destined to be overturned; we are constantly revising our understanding of ourselves as well as of our world. It would be difficult not to see how our understanding of metamorphosis has come a long way, from spontaneous generation to *deus sive natura* to the molecular revolution. But as Rumi recognized, our transformations are forms of return to the self:

> *We began*
> *as a mineral. We emerged into plant life*
> *and into the animal state, and then into being human*
> *and always we have forgotten our former states,*
> *except in early spring when we slightly recall*
> *being green again.*
>
> *That's how a young person turns*
> *towards a teacher. That's how a baby leans*
> *towards the breast, without knowing the secret*
> *of its desire, yet turning instinctively.*
>
> *Humankind is being led along an evolving course,*
> *through this migration of intelligences,*
> *and though we seem to be sleeping,*
> *there is an inner wakefulness*
> *that directs the dream,*
> *and that will eventually startle us back*
> *to the truth of who we are.*

Sol stares at me, smiling. For a merciful moment, I see the world through her eyes. Then I am myself again. I think: Rumi's words sound like a hallucination. A vision only few will ever achieve. But maybe Rumi remains loved, across oceans and lands and ages, because he saw a truth that resonates with many of us: We return to ourselves, perennially, but like Nabokov's Aurelian and Strauss's pining strings, never quite reach our destination.

"Yesterday I was clever, so I wanted to change the world," Rumi wrote toward the end of his life. "Today I am wise, so I am changing myself."

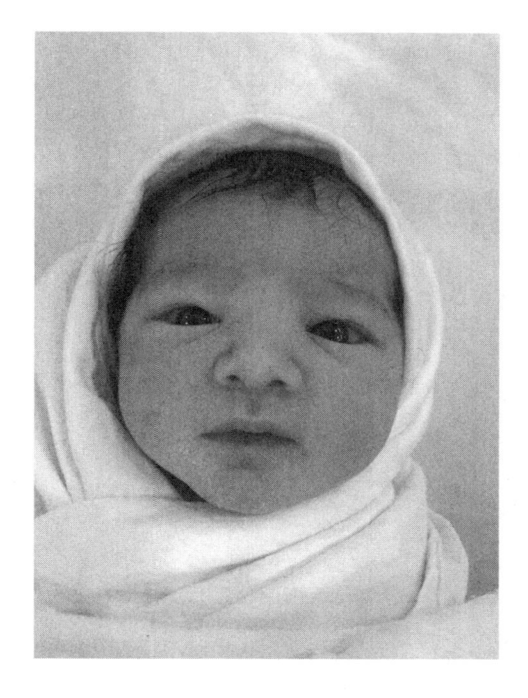

Baby Sol.
Photo: Oren Harman.

ACKNOWLEDGMENTS

I made my first attempt at this book during that strange COVID 19–dominated period, as if in a white cloud. It had a different title then—*Metamorphosis: A Natural Wonder, an Evolutionary Mystery, and a Father's 40-Week Journey to Understand the Deep Meaning of Transformation and Change*—but I'd failed to do my subject justice. I put it aside for some time to gain more clarity. My family and I moved to Berlin, where I spent a year thinking about metamorphosis beside a beautiful lake at the Institute for Advanced Study. That year, 2023, my father died in Jerusalem. Some months later I joined the insect immunology lab of Jens Rolff at the Freie University for twelve months, and learned an enormous amount about physiology and evolution. I also joined Katja Krause's research group at the Max Planck Institute for the History of Science, where I learned how to look more closely at a text. I had the opportunity to travel to Maria Sibylla Merian's and Haeckel's old homes and archives, and to visit Lynn Riddiford and Jim Truman on their island. I spent long days and nights in an ashram in India. The time away, and the experiences it brought with it, transformed the book entirely. It also changed our family in ways that make me smile.

I have so many wonderful people to thank for this journey. Katja for discussions about Aristotle, Jens for arguments about the larva and pupa, and both for their kind generosity, German versions of Bedouin hospitality. To Janet Browne and Alexandre Metraux for their friendship and initial ideas on the subject. To Jimena Canales for reminding me of the hookah, Rose Eleanor O'Dea for the ping-pong,

Martin and Miriam Saar for the fox, and Daven Presgraves for all the ideas we shared and for the times you choked up. To Dieter Ebert for his approach to me at the Weizmann Institute, and for his thoughts on the metamorphic continuum. Thanks to Bernie Avishai for a walk on the old Jerusalem–Beirut train tracks, to Lorraine Daston for her wisdom, to Alan Lightman and Peter Galison for their art. To my friends at Amma's ashram in Amritapuri, Kerela—Bhavani Rao and Swami Achyutamrita Chaitanya—for deep conversations about the life cycle and transformations. To Daniel Knebel, my sweet and wise officemate, now a father himself. To Olivia Judson for our shared passions, Tuesday lunches, and nature walks in the Tiergarten. Thanks to the many archivists who helped me so generously, and the wonderful librarians at the Wiko and MPI. To Camillo Barbosa for introducing me to Jens, to Jens for introducing me to Stuart Reynolds, and to Stuart for sharing his wisdom and for introducing me to Jim Truman and Lynn Riddiford. Thank you, Jim and Lynn, from the bottom of my heart for opening your lab and home to me, and for indulging my every question. Your dedication to one another and your love for science are truly inspiring.

Thanks to the music historian Jeremy Eichler for discussions about Strauss, the writer Maria Stepanova for discussions about Nabokov, the late and great Daniel Tsadik for discussions of Averroes and Rumi, my classicist friend Ariadne Konstantinos for her thoughts on Ovid and his nymphs. Thanks to the lovely Inna Gertsberg for her drawings, and to Ofer Ben Zvi, a true Peter Pan, for his underwater camera. Thanks Andrea, Vera, Nina, Maike, Katharina, and Dunia not least of all. Thank you, longtime director of Haeckel's Phyletic Museum in Jena, Martin Fisher; Gerit Linneweber for thoughts on finding individuality in nervous system architecture; Janina Wellman for your attention to images; Moshe Bar and Yanay Ofran for being there when it counted most; and the two Ohads, Nachtomy and Parnes. Thank you to the many people who shouldered my nudging, answered my questions, or were kind enough to share their wisdom: Mary and Fred Nijhout, Xavier Bellés, Eva Jablonka, Giuseppe Testa, Lauren Redniss, Shlompi

Peled, Uri Nitzan, Gilad Malach, Yoni Turner, Sigalit Landau, Oded Rechavi, Shimon Schocken, Igal Myrtenbaum, Salit Kark, Ben Reis, David Plunkett, Ghil'ad Zuckermann, Yehuda Melzer, Yoav Beirach, Maximilian Benz, Lynda Delph, Curtis Lively, Ittai Weinryb, Milica Nikolic, Tchavdar Marinov, Elcin Aktoprak, Gunnar Hindrichs, Goggy Davidovich, Thomas Kaufmann, Kateryna Mischenko, Antonin Pottier, Lucia Ronchetti, André Schneider, Michael and Barbara Taborsky, Sultan Sooud al-Qassemi, Alexandre Courtiol, Uli Steiner, Sophie Armitage, Dirk Mikolajewski, Jonathan Jeschke, Jill Leutgeb, Adam Wilkins, Raghavendra Gadagkar, Giovanni Galizia, Cristoph Möllers, Dieter Grimm, The Shamna, Mark Borrello, David Sepkoski, Michael Dietrich, Hans-Jörg Rheinberger, Daniel Kevles, Erez Levanon, Fred Tauber, Shai Lavi. Thank you, Onur Erk Kavlak, for reintroducing me to insect dissection; and to Natalia Konopińska and Mariam Keshavarz for trying to teach me how to perform a proper allatectomy (I swear, I'll never kill another insect!). To Aryeh Tzaban of Bar Ilan University for affording me time, and to my colleagues there for their abiding patience. To the entire Van Leer Jerusalem Institute and Wiko staffs, thanks for beings homes of warmth and generosity.

Thanks to my gracious agents, Sarah Chalfant and Rebecca Nagel. To my kind editors, T. J. Kelleher at Basic Books in America and Georgina Laycock at John Murray in England. To Gillian Sutliff and Annie Chatham, and Dana Henricks, for her eagle-eyed, humor-filled copyediting. A special thanks to Lorin Stein, whose wisdom and kindness made all the difference, and to my friend Doron Weber, and the Alfred P. Sloan Foundation, for believing in me, yet a second time.

Above all, thanks to my family. My loving mother, Dorothy, my funny sister, Danna, my loyal brother, Mishy, my late father and compass, David. Thank you, Yaeli, the mother of this book and the mother of our children, for being exactly who you are, and accepting my transformations. Thanks, Shaizee, Abie, and Sol, my greatest teachers of all—for your laughter and tears and mischief. We have one ride in this life, one gilgul—that's all. I can't believe I get to share mine with you.

SOURCES

Preface: Toads get a bad rap as disgusting and ugly, a test that kissing princesses must endure to find true love. But leave it to Orwell to see their striking beauty. His wonderful description quoted in the text comes from an essay he wrote in 1946 called "Some Thoughts on the Common Toad," in which presentiments of *1984*, like a toad in April, are beginning to stir. It concludes with the words: "The atom bombs are piling up in the factories, the police are prowling through the cities, the lies are streaming from the loudspeakers, but the earth is still going round the sun." As it happens, he chose an apt example. After spadefoot toads lay their eggs in ponds made by spring rains in the desert, the tadpoles that hatch in spacious ponds eat algae until they develop into juveniles and bury themselves in the sand. But the ones that just happen to be born into cramped ponds turn out differently, developing wide-opening jaws. To survive they eat each other, so that they can grow more quickly before the shallow pond dries up. Both forms of toad have the exact same genes, but only big ponds make gentle burrowers. Little ponds, alas, first produce cannibals.

Because we are constantly changing, when we make dramatic choices in life, like having children, we can't really know how they'll impact our future selves. In her book *Transformative Experience* (Oxford University Press, 2015), the philosopher L. A. Paul argues beautifully that such choices should be viewed as discoveries about the intrinsic nature of experience.

The Ovid quote "How many creatures walking on this earth / Have their first being in another form?" comes from the preface to his classic *Metamorphoses*, in the plural, which was written in the year 8 of the Common Era, contains fifteen books and 11,995 lines, and inspires every page of this book.

Part One: "Where Do We Come From?"

1. "Surinam": A beautifully written popular biography of Maria Sibylla Merian is Kim Todd's *Chrysalis: Maria Sibylla Merian and the Secrets of Metamorphosis* (Houghton Mifflin Harcourt, 2007), which served as one of the main inspirations and sources of materials for the sections on Merian in Part 1, and from which quotes of Merian in English come, unless otherwise indicated. Further sources include the more recent and scholarly book by Kay Etheridge, *The Flowering of Ecology: Maria Sibylla Merian's Caterpillar Book*, with translations from the original German texts by Michael Ritterson (Brill, 2021), as well as her earlier essay "Maria Sibylla Merian: The First Ecologist?," in *Women and Science: Figures and Representations—17th Century to Present*, edited by V. Molinari and D. Andreolle (Cambridge Scholars Publishing, 2011). Also, Kurt Wettengl's *Maria Sibylla Merian, Artist and Naturalist* (G. Hatje, 1998), as well as Natalie Zemon Davis, *Women on the Margins: Three Seventeenth-Century Lives* (Harvard, 1995), and specifically William T. Stern, "The Plants, the Insects and Other Animals of Merian's Metamorphosis Insectorum Surinamensium," in *Metamorphosis Insectorum Surinamensium* (Pion Limited, 1980). For younger readers there are two illustrated children's books, both award winners: *Maria Sibylla Merian: Artist, Scientist, Adventurer*, by Sarah B. Pomeroy and Jeyaraney Kathirithamby (Getty Publication, 2018), and Joyce Sidman's *The Girl Who Drew Butterflies: How Maria Merian's Art Changed Science* (Houghton Mifflin Harcourt Books for Young Readers, 2018). On the colonial history of Surinam, see Cornelis Goslinga, *The Dutch in the Caribbean and on the Wild Coast, 1580–1680* (University of Florida Press, 1971), as well as James Rodway, *Guiana: British, Dutch, and French* (T. Fisher Unwin, 1912), and the fascinating *Narrative of the Five Years Expedition Against the Revolted Negroes of Surinam*, by John Gabriel Stedman (Johns Hopkins University Press, 1988). For early adventures in

Surinam, see Garcilaso de la Vega and others, *Expeditions into the Valley of the Amazons, 1539, 1540, 1639* (London: Hakluyt Society, 1859), and Mark Herman, *Searching for El Dorado* (Doubleday, 2003). On Sir Walter Raleigh, see Stephen J. Greenblatt, *Sir Walter Raleigh: The Renaissance Man and His Roles* (Yale, 1973). On nature in Suriname, see *Natural History and Ecology of Suriname*, edited by Bart De Dijn (LM Publishers, 2018). The first book of Suriname history written by a Surinamese man is Anton de Kom's excellent *We Slaves of Suriname* (Polity, 2022).

2. "Cicada": There are a number of wonderful books about cicadas, including the quaintly dated *Insect Singers: A Natural History of the Cicadas*, by J. G. Myers, from 1929, as well as Shaun Tan's 2019 illustrated work of fiction *Cicada*, about a hardworking but underappreciated cicada toiling away in an office. Quotes from the ancients, including Antipater, appear in F. C. Clark, "Song of the Cicada," *American Naturalist* 9, no. 2 (1875): 70–74. On our current knowledge about cicadas in Suriname, see Allen F. Sanborn, "The Cicadas (Hemiptera: Cicadidae) of Suriname Including the Description of Two New Species, Five New Combinations, and Three New Records," *Zootaxa* 4881, no. 3 (2020). For an early mathematical model of synchronized cicada emergence, see Frank C. Hoppensteadt and Joseph B. Keller, "Synchronization of Periodical Cicada Emergences," *Science* 194, no. 4262 (1976): 335–337, and for a more recent study of the evolution of periodical emergences see Y. Tanaka, J. Yoshimura, C. Simon, J. Cooley, and K. Tainaka, "The Allee Effect in the Selection for Prime-Numbered Cycles in Periodical Cicadas," *Proceedings of the National Academy of Sciences* 106, no. 22 (2009): 8975–8979. A particularly charming encounter between Sir David Attenborough and a horny North American male *Magicicada* can be found in this YouTube video from his BBC film *Life in the Undergrowth*: https://www.youtube.com/watch?v=tjLiWy2nT7U.

3. "Aristotle": A freshly informative, though somewhat whiggish, book rehabilitating Aristotle as an important father of science, and biology in particular, is *The Lagoon: How Aristotle Invented Science* (Viking, 2014) by Armand Marie Leroi. Contrast his account to the one produced by Ernst Mayr in his massive *The Growth of Biological Thought* (Harvard, 1982), and D'Arcy Wentworth Thompson's *On Aristotle as a Biologist*, which was the title of his Herbert Spencer Lecture at Oxford in 1913 and was published in 2018 by Forgotten Books; the latter two see Aristotle as a major stumbling block for the development of the modern scientific method. For a fuller explanation of Aristotle on metamorphosis and insects, read Stuart Reynolds, "Cooking Up the Perfect Insect: Aristotle's Transformational Idea About the Complete Metamorphosis of Insects," *Philosophical Transactions of the Royal Society B* 374 (2019). A scholarly explanation for why the belief in spontaneous generation lasted as long as it did can be found in Daryn Lehoux, *Creatures Born of Mud and Slime: The Wonder and Complexity of Spontaneous Generation* (Johns Hopkins University Press, 2017). Quotes from Aristotle's *Generation of Animals* are from the Loeb Classical Library Edition (1942), translated by Al Peck. All Ovid quotes come from his *Metamorphoses* (Penguin Classics, 2004), translated by David Raeburn. When it comes to Aristotle's notion of perfection, later commentators would use it for a moral purpose: applying a threefold division of the sciences into rational sciences (logic); real sciences (mathematics, natural sciences, metaphysics); and practical sciences (ethics and politics). The Dominican thinker Albertus Magnus of Cologne argued that by studying them in order, a person could reach the *telos* of *perfectio* of a fully realized human being. The purpose of ethics, on this hierarchical ordering of knowledge, is to perfect the human being. I wonder how many politicians today would agree!

4. "Imposters": Readers will be glad to know that not all butterfly-ant relations are deceitful: *Apharitis cilissa,* a butterfly endemic to Syria, Iran, Iraq, and Israel, lays its eggs near ant colonies, and its larvae, just like those of the large blue, are tended to by ants. Unlike the large blue, *Apharitis* is a mutualist: Its larvae exude a sweet liquid that nourishes the ant larvae. The large blue, for its part, remains on the IUCN Red List of Threatened Species, and has been studied extensively in order to assist conservation and preservation efforts. Had scientists known earlier about its unique life cycle, perhaps intervention may have helped—an argument for Maria Sibylla Merian's ecological approach. On early attempts to understand why its numbers were dwindling, and early attempts at conservation, see J. A. Thomas, "Why Did the Large Blue Become Extinct in Britain?," *Oryx* 15, no. 3 (1980): 243–247, and *Rare Species Conservation: Case Studies of European Butterflies* (Blackwell Scientific Publication, 1991). On the ways in which the large blue pinpoints specific species of ant to parasitize, read Magdalena Witek et al., "Host and Specificity of Large Blue Butterflies *Phengaris* (*Maculinia*) (Lepidoptera: Lycaenidae) Inhabiting Humid Grasslands in East-central Europe," *European Journal of Entomology* 105, no. 5 (2008): 871–877, and on manipulating chemical cues for purposes of conservation see R. M. Guillem, "Using Chemo-taxonomy of Host Ants to Help Conserve the Large Blue Butterfly," *Biological Conservation* 148 (2012): 39–43. On its amazing acoustical trickery, see F. Barbero, J. A. Thomas, S. Bonelli, E. Balletto, and K. Schonrogge, "Queen Ants Make Distinctive Sounds That Are Mimicked by a Butterfly Social Parasite," *Science* 323, no. 5915 (2009): 782–785. Finally, on the "queen effect," see J. A. Thomas and J. C. Wardlaw, "The Effect of Queen Ants on the Survival of Maculinae arion Larvae in Myrmica Ant Nests," *Oecologia* 85, no. 1 (1990): 87–91. An amazing YouTube video showing the entire life cycle of large blues, including their host ants and wasp predators, can be seen in the BBC film *Life in the Undergrowth*, once again with the tireless and charming Sir David Attenborough: https://www.youtube.com/watch?v=GCo2uCLXvhk.

Great biographies have been written about Catherine the Great, of which Robert K. Massie's *Catherine the Great: Portrait of a Woman* (Random House, 2012); and Virginia Rounding's *Catherine the Great* (Hutchinson, 2006) served me well, alongside *The Memoirs of Catherine the Great*, translated by Markus Cruse and Hilde Hoogenboom (Modern Library Classics, 2006). The quip about Catherine's musical abilities comes from Sophy Roberts, *The Lost Pianos of Siberia* (Penguin, 2020). Two recent television shows help spark the imagination: the HBO historical series *Catherine the Great*, starring Helen Mirren, which focuses on her later years, and the fictionalized, satirical *The Great*, from Hulu Originals, starring Elle Fanning, concentrating in its first season on Sophie/Catherine's path to the throne. Both are recommended!

5. "The Transfiguration of Jesus to Christ": Maria Sibylla Merian left very little after her in writing. It is very difficult to glean personal details from her scientific notebooks, or from the eighteen letters that survive, mainly technical and dry. This makes trying to get into her head and conjure her voice all the more of a challenge; exercising some literary freedom, I did my best to stay loyal to known facts. In particular, I was helped by the brilliant translation by Michael Ritterson of Merian's *Raupen* book, and the notes that went into producing it, in Kay Etheridge, *The Flowering of Ecology*. Also useful was Elisabeth Rucker's "The Life and Personality of Maria Sibylla Merian," in *Metamorphosis Insectorum Surinamensium* (Pion Limited, 1980), as well as Kim Todd's wonderful book. Besides the sources mentioned under the "Surinam" chapter, I have consulted Merian's original printed works, and their translations, six books in all, two published posthumously. Particularly helpful were the scientific notebooks, housed in the library of the Academy of Sciences in Leningrad and reprinted as Maria Sibylla Merian and Wolfgang Dietrich Beer, *Maria Sibylla Merian: Shmetterlinge, Käfer und andere Insekten. Leningrader Studienbuch* [*Books of Notes and Studies*] (Reich, 1976). After paying a visit to Nuremberg, to the Graff home, which miraculously survived the Allied bombing of the city in World War II, as well as to St. Sebald's Church, the Albrecht Dürer house, and the flower and cherry-tree garden beneath the old imperial castle where Sibylla Merian would collect eggs and caterpillars and butterflies, I looked at her wills and legal actions at the *Institut für Stadtgeschichte*

Frankfurt am Main and the *Gemeentearchief* Amsterdam, and at the eighteen letters (scattered in various collections at the British Library, University of Erlangen, Fondation Custodia in Paris, *Stadtbibliothek* Nuremberg, and the *Germanisches Nationalmuseum* Nuremberg). I have also "read" extensively in the period, to try to catch words and phrases and ways of expression. Secondary literature that was especially helpful for this section includes Elizabeth L. Eisenstein, *The Printing Revolution as an Agent of Change* (Cambridge, 1979); Eric Jorink, *Reading the Book of Nature in the Dutch Golden Age, 1575–1715* (Brill, 2010); and Pamela Smith, *The Body of the Artisan: Art and Experience in the Scientific Revolution* (Chicago, 2004).

6. *"Ex Ovo Omnia?"*: For centuries Kircher was derided as "the man who got everything wrong," but he's enjoying a reappraisal in recent years, not least for his prescience with regard to the connection between disease and microorganisms. See, for example, the volume that came out of the 2002 New York Institute conference declaring him to be "the coolest man who ever lived," *Athanasius Kircher: The Last Man Who Knew Everything*, edited by Paula Findlen (Routledge, 2004). Athanasius Kircher's own whale of a book was called *Mundus Subterraneus* (Amsterdam: Waesberge, 1665); the quote comes from a translation by P. Gottdenker in "Francesco Redi and the Fly Experiments," *Bulletin of the History of Medicine* 53 (1979): 575–592. It seems clear that Redi was responding to the section in Kircher's book which dealt with the generation of insects. Redi began his experiments either in the summer of 1666 or 1667, and his book *Esperienze Intorno alla generazione degl'insetti* (*Experiments on the Generation of Insects*) appeared in 1688; quotes taken from the English translation from 1909 from Open Court, as well as Redi's earlier treatise on viper poison, *Osservazioni intorno alle vipere*, translated by P. K. Kneopfel as *Francesco Redi on Vipers* (Brill, 1988). The real name of the Swiss occultist, physician, alchemist, and philosopher Paracelsus was Philippus Aureolus Theophrastus Bombastus von Hohenheim, and he provided the recipe for a tiny human in his book *Of the Nature of Things* from 1537. As much as it might sound counterintuitive, he actually played a role moving away from bookish knowledge toward experiment: See Walter Pagels, *Paracelsus: An Introduction to Philosophical Medicine in the Era of the Renaissance*, 2nd revised edition (S. Karger, 1982). On John Wilkins's natural language, see his *An Essay Towards a Real Character*, and a *Philosophical Language* from 1668, and another super-interesting source by John Bulwer, besides his book *Anthropometamorphosis*, is his work on the natural language of the body called *Pathomyatomia* from 1649, which studies the actions of the muscles of the face to produce expressions. Information on Harvey, including the quote about his character, I gleaned from D'arcy Power's 1897 biography, *William Harvey: Masters of Medicine* from T. Fisher Unwin. Extremely helpful to me in getting the big picture was Matthew Cobb's excellent book *Generation: The Seventeenth-Century Scientists Who Unraveled the Secrets of Sex, Life and Growth* (Bloomsbury, 2006), as well as Lorraine Daston and Katherine Park, *Wonders and the Order of Nature 1150–1750* (Zone Books, 2001).

7. "An Ingenious Woman": Merian noted that she first began observing moths and butterflies in 1660. Nearly twenty years of such study would have gone into her caterpillar book from 1679, whose short title was *Der Raupen wunderbare Verwandelung*. One book that she attests had a particular impact on her is Johannes Goedaert, *Metamorphosis et Historia Naturalis Insectorum*, 3 volumes (Jacobum Fiernsiuem, 1662–1668). Interestingly, compared to Goedaert, Merian used less overtly religious language, hinting that Nature on its own terms interested her more than God's nature. Still, she was a woman of her times and would not have been untouched by the tradition of studying nature as a way to study God ("Therefore, seek herein, not mine but God's glory alone, and glorify Him as creator of even the smallest and least of worms," she writes in her introduction—though this surely was also insurance against those railing against wasting efforts on "lowly" creatures). Merian would not have been the only woman to paint flowers or insects in her day, and women in liberal Frankfurt and Nuremberg were afforded recognition and respect in the various trades, including still-art painting, natural history, and publishing. Indeed, idleness of any kind and by any gender was frowned upon in this Lutheran city. There was a large community of both men and women who engaged in the arts, literature, art collecting, and the pursuit of nature—see Sabina Leßman, "Susana Maria von Sandrart: Women Artists in 17th Century Nürnberg," *Woman's Art Journal* 14, no. 1 (1993); and Elizabeth Alice Honig, "The Art of Being 'Artistic': Dutch Women's Creative Practices in the 17th Century," *Woman's Art Journal* 22, no. 2 (2001). Also very useful to me was a PhD dissertation by Tomomi Kinukawa of the University of Wisconsin at Madison titled *Art Competes with Nature: Maria Sibylla Merian (1647–1717) and the Culture of Natural History*; as was Ad Stijnman, *Engraving and Etching: 1400–2000* (Archetype Books, 2012), on the intricacies of etching, engraving, plate making, and painting. Still, women were not allowed to learn at the art academies, nor join the guilds, and would need to be enterprising about earning a living. See Sheilagh Ogilvie, *A Bitter Living: Women, Markets, and Social Capital in early Modern Germany* (Oxford, 2003); and also Heide Wunder, *She Is the Sun, He Is the Moon: Women in Early Modern Germany*, translated by Thomas Dunlap (Harvard, 1998). As her notes and writings attest, Merian had both ambition and business acumen. She hoped to gain the "notice of posterity," and was pleased, she wrote, to "earn the praise and favor of great men."

8. "Gentlemen": The story of the feuds between the men who cracked the mystery of generation in the second half of the seventeenth century is told in many places, but nowhere more clearly than in Matthew Cobb's aforementioned book *Generation*, from which English translations of the quotes come, unless otherwise indicated. The original primary sources I consulted include Niels Steno, *Elementorum Myologiae Specimen* (Florence: Stella, 1668); Reinier de Graaf, *De Vivorum Organis Generationi Inservientibus, de Clysteribus et de usu Siphonis in Anatomia* (Leiden: Hack, 1668), and *De Mulierum Organis Generationi Inservientibus* (Leiden: Hack, 1672); Jan Swammerdam, *Disputatio Medica Inauguralis, Continens Selectas de Respiratione*

Propositiones (Leisen: Elsevier, 1667), and *Historia Insectorum Generalis* (Utrecht: van Dreunen, 1669), as well as the posthumous *Bybel der Natuur* (Amsterdam: Herman Boerhaave, 1737), also published in English as *The Book of Nature* (London: Seyffert, 1758); Marcellius Malpighi, *Dissertatio Epistolica de Bombyce* (London: Martin & Allestry, 1669); Theodor Kerckling, *Anthropogeniae Ichonographia, Sive Conformatio Foetus ab Ovo* (Amsterdam: n.p., 1671). On the doctrine of preformation, I was informed by Peter Bowler, "Preformation and Pre-existence in the Seventeenth Century," *Journal of the History of Biology* 4, no. 2 (1971): 221–244. An enjoyable popular book on the life of Steno focusing on his geology is Alan Cutler's *The Seashell on the Mountaintop* (Dutton, 2003). As for Swammerdam, while remembered by historians of science both as a pioneer of the study of insects' female and male sexual anatomy, and developer of techniques to study bodies such as wax injection into blood vessels and a method for the preparation of hollow human organs, he is also remembered as a herald of the natural theology of the eighteenth and nineteenth centuries, in which God's design was detected everywhere from the crystalline eye of the fish to the movement of the stars in the heavens.

9. "White Witch": Anna Maria van Schurman's book exhorting women to enter the worlds of men was translated and can be read with great delight: *Whether a Christian Woman Should Be Educated and Other Writings from Her Intellectual Circle*, edited and translated by Joyce I. Irwin (University of Chicago Press, 1998). William Penn's account of his visit to the Labadist colony in Wiuwert is detailed in "Travels in Holland and Germany" in *The Selected Works of William Penn*, vol. 2 (Kraus Reprint Company, 1825). Petrus Dittlebach's scandalous tell-all book about the Labadist community in Wiuwert was published in Dutch in 1692, a year after Maria left, and was called *Verval en val der Labadisten*. The book became a bestseller and did not shy away from detailing jealousies, heresies, and other human foibles. But it was a mixture of disease, spiritual defeat, and the desire to return to a normal life that ultimately defeated many of the sect. For more on Labadie and his sect, see T. J. Saxby, *The Quest for the New*

Jerusalem: John de Labadie and the Labadists, 1610–1744 (Springer, 1987). Maria Sibylla Merian's gorgeous Surinam book is called *Metamorphosis Insectorum Surinamensium* (Amsterdam, 1705), and if you have an opportunity to gaze at an original, I highly recommend that you do. Current research at the Center for Scientific Studies in the Arts at Northwestern University employs hyperspectral imaging and macro X-ray fluorescence to study the elemental makeup of the pigments in the book, showing (perhaps disappointingly) that many prints have been recolored using cadmium red, titanium white, and Prussian blue unavailable in Merian's time.

10. "Nabokov": "The Aurelian" was written by Nabokov in Russian while he was in exile in Berlin, in 1930. Together with Peter Pertzov, he translated it to English, and it was published in *The Atlantic Monthly* in 1941. For more on Nabokov's life, I recommend his beautiful memoire, *Speak, Memory* (Vintage, 1989), alongside Brian Boyd, *Vladimir Nabokov: The Russian Years*, and *Vladimir Nabokov: The American Years*, both from Princeton University Press. More recently, Andrea Pitzer examines his writings and politics incisively against the backdrop of a turbulent twentieth century in *The Secret History of Vladimir Nabokov* (Pegasus, 2014), and a wonderful new collection of Nabokov writings provides further penetrating insights: *Think, Write, Speak: Uncollected Essays, Reviews, Interviews and Letters to the Editor*, edited by Brian Boyd and Anastasia Tolstoy (Penguin, 2019), as does an interview published in the *Paris Review* in 1967, "Vladimir Nabokov, The Art of Fiction No. 40," from which the quotes come. Much has been written about Nabokov the entomologist: K. Johnson, G. W. Whitaker, and Z. Balint's "Nabokov as Lepidopterist: An Informed Aappraisal," *Nabokov Studies* 3 (1996): 123–144; Kurt Johnson and Steve Coates's *Nabokov's Blues: The Scientific Odyssey of a Literary Genius* (Zoland Books, 2000); and Dieter E. Zimmer's *A Guide to Nabokov's Butterflies and Moths*, from 2001, for example. While he did accept that evolution occurs, Nabokov didn't believe natural selection could explain mimicry in butterflies (nor did he think genetics was at all useful for taxonomic purposes); for more on this, see Victoria N. Alexander, "Nabokov, Teleology and Insect Mimicry," *Nabokov Studies* 7 (2002/2003): 177–213, which to my mind gives Nabokov's arguments more credit than they deserve. An insightful essay on Nabokov as observer remains Stephen Jay Gould's "No Science Without Fancy, No Art Without Facts: The Lepidoptery of Vladimir Nabokov," in his book *I Have Landed: The End of a Beginning in Natural History* (Harmony Books, 2002). Readers may also be interested in Peter Medak's short film *Nabokov on Kafka*, starring the late Christopher Plummer and filmed in 1985, a dramatization of Nabokov's lectures on Franz Kafka's *The Metamorphosis* available on YouTube. As for Antonie van Leeuwenhoek's "private affair" in discovering the contents of sperm, the drama can be read about in the reproduction of his letter to the Royal Society in F. J. Cole, *Early Theories of Sexual Generation* (Oxford, 1930), from which the quotes come. More on the "ovist" versus "spermist" debates, and how they were ultimately solved, has been written about nicely in Clara Pinto-Corriera, *The Ovary of Eve: Egg and Sperm and Preformation* (Chicago, 1997), which also tells the story of the use and misuse of Hartsoeker's homunculus image over time. The reason why the seventeenth century saw a return to the doctrine of preformation was because Harvey's epigenesis depended on a vital factor present in the semen. The new microscopists rejected vitalism, believing that it led to occultism, or even atheism. They sought a balance between intelligible material mechanicism and godly natural order instead. As for Linnaeus, he has received many appreciations. I recommend a new biography by Gunnar Broberg, *The Man Who Organized Nature: The Life of Linnaeus* (Princeton, 2023). For those who want to understand how the science of taxonomy has changed from the days of Linnaeus, read Carol Kaesuk Yoon's superb *Naming Nature: The Clash Between Instinct and Science* (W. W. Norton, 2010). Finally, on the evolution and vicissitudes of Maria Sibylla Merian's reputation, see Chapter 8 of the aforementioned book by Kim Todd, *Chrysalis*. An essay that did particular damage, pointing out genuine mistakes but also doing her many injustices, was written by a young reverend named Lansdown Guilding, from the Caribbean island of St. Vincent. It appeared in the

Magazine of Natural History in 1834 under the title "Observations of the work of Maria Sibylla Merian on the Insects, &c., of Surinam."

11. "Mayfly": The mayfly is a generic name for up-wing members of one of the oldest orders of insects, the Ephemeroptera. There are about three thousand known species of mayfly in this order, which includes dragonflies and damselflies. As its name suggests, the mayfly used to emerge exclusively in the month of May, but due to a discrepancy of eleven days caused by the change from the Julian to Gregorian calendar, the adults now sometimes arrive in June as well. During its immature aquatic stage, the mayfly is called a naiad, which in ancient Greek folklore denoted one of the four groups of nymphs, the Meliae (ash tree nymphs), the Naiads (freshwater nymphs), the Nereids (sea nymphs), and the Oreads (mountain nymphs). Already in classical times, both Aristotle and Pliny the Elder noted the brief lives of mayfly adults. In 1495 the German artist Albrecht Dürer produced an engraving called *The Holy Family with the Mayfly* to suggest a link between heaven and earth. Nike produced a line of running shoes in 2003 called "Mayfly" that were said to have a finite lifetime like their namesake. For more on the mayfly, go to the website Mayfly Central at Purdue University: https://www.entm.purdue.edu/mayfly/. "How to Survive a Massive Mayfly Storm," from the Entomology Today website of the Entomological Society of America, may also be useful. It has a video showing just how dramatic their taking to the air every May really is.

Part Two: "Where Are We Going?"

1. "Anna": A beautifully written scholarly biography of Haeckel is Robert J. Richards, *The Tragic Sense of Life: Ernst Haeckel and the Struggle over Evolutionary Thought* (University of Chicago Press, 2008), which informs Part 2 of the book throughout, and which is the source of English translations of Haeckel quotes, unless indicated otherwise. Also helpful to me in getting a multifaceted picture of Haeckel were: E. S. Russell's *Form and Function: A Contribution to the History of Animal Morphology* (1916; reproduced by the

University of Chicago Press, 1982); Gavin de Beer, *Embryos and Ancestors* (Oxford, 1940); Daniel Gasman, *The Scientific Origins of National Socialism: Social Darwinism in Ernst Haeckel and the German Monist League* (New York: Science History Publications, 1971); Stephen Jay Gould, *Ontogeny and Phylogeny* (Harvard, 1977); and Peter Bowler, *The Non-Darwinian Revolution* (Johns Hopkins University Press, 1988). In his brilliant biography, Richards set out to revive Haeckel's reputation, following three generations of critical appraisal both of his science and his politics. Richards was arguing expressly against Russell, de Beer, Gasman, Gould, and Bowler. Sander Joel Gliboff's PhD thesis, "The Pebble and the Planet: Paul Kammerer, Ernst Haeckel, and the Meaning of Darwinism," submitted to Johns Hopkins University in 2001, provides a further angle. I tried in the sections on Haeckel that follow to conjure what I think is the closest portrayal of him—the scientist, artist, thinker, and man—based on the vast Haeckel corpus, and especially primary sources. In this section, the ones most valuable to me have been Haeckel's *Der Kampf um den Entwickelungs-Gedanken: Drei Vorträge, gehalten am 14, 16 und 19 April 1905 in Saale der Sing-Akademie zu Berlin* (Berlin: Georg Reimer, 1905); his letters to Anna in *Himmelhoch Jauchzend: Erinnerungen und Briefe der liebe*, edited by Heinrich Schmidt (Dresden: Carl Reissner, 1927); and further letters to Anna housed in the Haeckel-Haus Collection at Jena and those printed in Haeckel's *Italienfahrt: Briefe an die Braut, 1859–1860*, edited by Heinrich Schmidt (Liepzig: K. F. Koeler, 1921). On Haeckel's special friendship and experiences with the poet and painter Hermann Allmers whom he met during his trip to Italy, see *Haeckel und Allmers: Die Geschichte einer Freundschaft in Briefen der Freunde*, edited by Rudolph Koop (Brememn: Arthur Geist, 1941). On his teacher's secular philosophy of science, see Rudolph Virchow, *Die Gründung der Berliner Universität und der Übergang aus dem philosophischen in das naturwissenschaftsliche Zeitalter* (Berlin, 1893). Haeckel speaks of his relationship with his teacher Müller, whom he thought committed suicide, both in the aforementioned *Italienfhart* and *Himmelhoch Jauchzend*, but a good appreciation of Müller's important influence can be found in Ruth G. Rinard, "The Problem of the Organic Individual: Ernst Haeckel and the Development of the Biogenetic Law," *Journal of the History of Biology* 14, no. 2 (1981): 249–275; as well as in Laura Otis, *Müller's Lab* (Oxford, 2007), which tells the story of seven of Johannes Müller's famous students, including Schwann, Virchow, and Haeckel, and the influence of the unusual man who trained them. Regarding Kant, the second part of his *Critique of Judgment* from 1790 contains a moral proof for the existence of God by way of introducing the concept of *intellectus archetypus*—a notion that would play a central role in shaping German idealism. Haeckel's poignant recollections of the medusa he found in Villafranca and named after his love exist tucked away in between technical descriptions in his whale of a book, *Das System der Medusen*, 2 vols. (Jena: Gustav Fischer, 1879). Finally, those who feel so moved are encouraged to visit Anna Sethe's grave in the quaint Johannisfriedhof, a historical cemetery near Jena city center, adjacent to the botanical garden. Her modest white marble tombstone rests quietly between vine leaves.

2. "Immortal Jellyfish": To learn more about the incredible *Turritopsis*, see the original article describing its life cycle: S. Piraino, F. Boero, B. Aeschbach, and V. Schmid, "Reversing the Life Cycle: Medusae Transforming into Polyps and Cell Transdifferentiation in *Turritopsis nutricula* (Cnidaria, Hydrozoa)," *Biological Bulletin* 90 (1996): 302–312 (Christian Sommer, the marine biology student who fished them out in Rapallo, was less quick to comprehend the significance of his discovery). On Kubota's attempts to understand the genetics of transdifferentiation, see Y. Hasegawa, T. Watanabe, M. Takazawa, O. Ohara, S. Kubota, "De Novo Assembly of the Transcriptome of *Turritopsis*, a Jellyfish That Repeatedly Rejuvenates," *Zoological Science* 33, no. 4 (2016): 366–371. On the promise of transdifferentiation in humans see Koji Tanabe et al., "Transdifferentiation of Human Adult Peripheral Blood T Cells into Neurons," *Proceedings of the National Academy of Sciences* 115, no. 25 (2018): 6470–6475. Kubota and colleagues have recently published a full genome analysis of *Turritopsis*, in Yoshinori Hagesawa et al., "Genome Assembly and

Transcriptomic Analyses of the Repeatedly Rejuvenating Jellyfish *Turritopsis dohrnii*," *DNA Research* 30, no. 1 (2022): 1–8.

Although there are few hydrozoan experts in the world, *Hydra* research has a history going back three hundred years and is a more active field of study than *Turritopsis*. In fact, the *Hydra* was the first animal in which asexual reproduction through budding, regeneration, and successful transplantation of tissues to other animals was shown: See Abraham Trembley, *Mémoires, pour servir à l'histoire d'un genre de polypes d'eau douce, à bras en forme de cornes* (Leiden: Jean and Herman Verbeek, 1744). For modern studies of immortality in *Hydra*, see Daniel E. Martínez's original article, "Mortality Patterns Suggest Lack of Senescence in *Hydra*," *Experimental Gerontology* 33, no. 3 (1998): 217–225, and a later study he coordinated with a collaborator in Germany, James W. Vaupel, showing this on a large sample: Ralf Schaible et al., "Constant Mortality and Fertility over Age in *Hydra*," *Proceedings of the National Academy of Sciences* 112, no. 51 (2015): 15701–15706. On the *Hydra* genome, a study for which Kevin Peterson did the miRNA work, see Jarrod A. Chapman et al., "The Dynamic Genome of *Hydra*," *Nature* 464 (2010): 592–596. Today, we know that a developmental gene called FoxO is an important regulator of stem-cell maintenance in immortal Hydra, safeguarding their indefinite self-renewal. See Anna-Marei Boehm et al., "FoxO Is a Critical Regulator of Stem Cell Maintenance in *Hydra*," *Proceedings of the National Academy of Sciences* 109, no. 48 (2012): 19697–19702, in which the researchers clearly show that down regulation of the gene leads to an increase in terminally differentiated cells, which results in a drastically reduced population growth rate. Compared to the usual workhorses *Drosophila* melanogaster and the nematode *C. elegans*, *Hydra* present an attractive model to help us understand the human process of aging: See Szymon Tomczyk et al., "Hydra, a Powerful Model for Aging Studies," *Invertebrate Reproduction and Development* 59, S1 (2015): 11–16. On the work of Erez Levanon regarding A-to-I RNA editing with ADAR enzymes, see Eli Eisenberg and Erez Y. Levanon, "A-to-I RNA Editing—Immune Protector and Transcriptome Diversifier," *Nature Reviews Genetics*

19 (2019): 473–490. Levanon and colleagues now plan to look at the role of this RNA editing system specifically in metamorphosis. The reference to Nietzsche and the quotes on "the silent invasion," "gelatin consciousness everlasting," and "we look like a damn jellyfish" come from a *New York Times Magazine* article by Nathaniel Rich from November 28, 2012, titled "Can a Jellyfish Unlock the Secret of Immortality?" Rich tells the story of the *Turritopsis* specialist, Shin Kubota, who can be watched explaining and singing about his beloved jellyfish, whom he calls "superheroes," here: https://www.youtube.com/watch?v=sTeer-apF0I and here: https://www.youtube.com/watch?v=o3Q7ETDlr3s.

3. "Goethe": In his day, Goethe's theory of plant development was considered by some to be revolutionary; they called him "the father of morphology," a word he himself had invented. Others saw Goethe as an amateur, his scientific musings idealistic and more like poetry. Only in recent decades has it become possible to test his intuitive theory through rigorous scientific experiment, allowing Goethe's ideas to be appreciated through a fresh new modern genetic perspective. And while some of his ideas have been deemed nonscientific, like the plant "purifying" its sap as it grows upward, incredibly, the basic insight of variations on a theme stands. So does the deep insight that abnormal development can be a key to understanding the normal way of things. For a modern rendering of plant and flower development through the gaze of Goethe's thought, see Marcello Carnier Dornias and Odair Dorenelas, "From Leaf to Flower: Revisiting Goethe's Concept of the 'Metamorphosis' of Plants," *Brazilian Journal of Plant Physiology* 17, no. 4 (2005): 335–343, as well as Enrico Coen and Rosemary Carpenter's "The Metamorphosis of Flowers," *The Plant Cell* 5, no. 10 (1993): 1175–1181. It's also worthwhile reading Coen's positive appraisal in "Goethe and the ABC Model of Plant Development," *Comptes Rendus de l'Acádemie des Sciences, Series III* 324, no. 6 (2001): 523–530. Helpful to me in understanding Goethe's scientific approach were Adolf Portmann, "Goethe and the Concept of Metamorphosis," in *Goethe and the Sciences: A Reappraisal*, edited by Frederick Armine, Francis J. Zucker, and Harvey Wheeler, Boston Studies in the Philosophy of Sciences, 97 (D. Reidel, 1987), 133–145; Lisbett Koerner, "Goethe's Botany," *Isis* 84 (1993): 470–495; Theresa M. Kelley, "Restless Romantic Plants: Goethe Meets Hegel," *European Romantic Review* 20, no. 2 (2009): 187–195; and in particular, Robert J. Richard, *The Romantic Conception of Life: Science and Philosophy in the Age of Goethe* (Chicago, 2002). A Goethean-inspired textbook is A. Shushantke, *Metamorphosis: Evolution in Action* (Adonis Press, 2009), which gives a feel for his phenomenological, and very unmodern, approach. I was also enlightened by the twentieth-century philosopher Paul Feyerabend's support of Goethe's challenge to Newton's theory of light, in Goethe's 1810 *Theory of Colours*, in which Feyerabend states that "Newton . . . did not give the explanation [of light] but simply re-described what he saw . . . [and] introduced the machinery of the very same theory he wanted to prove." Others have not found Goethe's approach as useful. Goethe's own writings were presented in translation by B. Mueller and C. J. Engard in *Goethe's Botanical Writings* (University of Hawaii Press, 1952), where the quote from the letter to Herder comes from (it appears, too, with the other Italian journey quotes included in the text, in his *Italian Journey*, available in paperback from Penguin). More recently, a beautiful new illustrated translation of *Metamorphosis of Plants* has become available with an introduction and photographs by Gordon L. Miller (MIT Press, 2009). For general readers, Richard Holmes's fabulous 2010 book *The Age of Wonder: The Romantic Generation and the Discovery of the Beauty and Terror of Science* is much recommended, especially on the English side of things. For a wonderful biography of Goethe, presenting him as a touchstone of the modern age, read Rüdinger Safranski's delightful *Goethe: Life as a Work of Art*, translated by David Dollenmayer (Liveright, 2018). Quotes about Goethe's philosophy of science come from his 1792 essay "The Experiment as Mediator Between Subject and Object," in *Scientific Studies* (*Goethe: The Collected Works*, vol. 12), edited by Douglas Miller (Princeton University Press, 1995). The quote about nature cursing stasis was used in the preface to Haeckel's own magnum opus, *Generelle Morphologie der Organismen* from 1866.

4. "Progress": The idea of progress has a fascinating history. I recommend starting with J. B. Bury's classic *The Idea of Progress: An Inquiry into Its Origin and Growth* (Blackmore and Dennett, 2018), and then reading Robert Nisbett, *History of the Idea of Progress* (Routledge, 1994). To learn more about the myriad ways in which the modern idea of Progress and our understandings of nature have interacted with and drawn from each other, especially through the prism of evolution, see Michael Ruse, *Monad to Man: The Concept of Progress in Evolutionary Biology* (Harvard University Press, 1997). Focusing on progress and Darwinism in the modern era, Haeckel's biographer Robert J. Richard provides a different perspective in "The Moral Foundations of the Idea of Evolutionary Progress: Darwin, Spencer, and the Neo-Darwinians," in *Evolutionary Progress*, edited by Matthew Nitecki (University of Chicago Press, 1988), 129–148. For a strong argument for the idea that we humans are, as the author writes, "a cosmic accident that would never arise again if the tree of life could be replanted," see Stephen Jay Gould, *Full House: The Spread of Excellence from Plato to Darwin* (Harmony, 1996). Conversely, for a controversial, but challenging, argument that there is purpose in the universe, see the philosopher Thomas Nagel's book *Mind and Cosmos: Why the Materialistic Neo-Darwinian Conception of Nature Is Most Certainly Wrong* (Oxford: Oxford University Press, 2012), and a critical review of the book by Eva Jablonka and Simona Ginsburg, "The Major Teleological Transitions in Evolution: Why the Materialistic View of Evolution Is Most Certainly Right," *Journal of Consciousness Studies* 20, nos. 9–10 (2013): 177–205. Finally, to experience how humankind's utopian dreaming has been changing over the ages, have a look at *The Utopia Reader*, edited by G. Claeys and L. T. Sargent (New York University Press, 1999). And to judge for yourselves whether this foundational modern text is a utopia or rather a thinly veiled dystopia, read Francis Bacon's *New Atlantis.* Both Darwin quotes on progress come from the same B Notebook from 1837, showing that his ambivalence about progress existed early on. Later and telling quotes from Darwin's letters on the question of progress to the American Lamarckian Alphaeus Hyatt and to the British Botanist Joseph Hooker can be found in the online Darwin Correspondence Project.

5. "The El Dorado of Zoology": Unlike Maria Sibylla Merian, who left very few personal sources behind her, Haeckel left many letters, travel books, and an impressive array of professional and public writings and speeches. Getting into his head and finding his voice is in a way almost an opposite problem to finding Merian's: He wrote so much, and in different registers, that one has to make judgments and take licenses in searching for an essence. I hope I have been able to do so with some justice, while sticking as closely as I can to the written record. The sources I've used in the upcoming Haeckel sections are varied: first, Haeckel's popular books, in particular (as translated into the English): *The History of Creation*, *The Evolution of Man*, *The Riddle of the Universe*, *The Wonders of Life*, *Freedom in Science and Teaching*, *Monism as Connecting Religion and Science*. Then the more technical monographs: *Radiolara*, *Siphonophora*, *Monera*, *Calcareous Sponges*; his technical (and most important book) *General Morphology of Organisms*, as well as *Systematic Phylogeny* and his extensive *Challenger* reports on Deep-Sea Medusae, Siphonophora, Deep-Sea Keratosa, and Radiolaria. Travel books include: *Travel Notes of India*, *Travel Notes of Malaysia*, *Art Forms in Nature*, *A Visit to Ceylon*. Haeckel's personal papers and correspondence are housed in Ernst Haeckel–Haus at the Friedrich-Schiller-Universität at Jena, and further materials exist in the (nachlaß) of Ernst Haeckel at the Staatsbibliothek in Berlin. There are also edited books, including *Haeckel-Korrespondenz: Übersucht über den briefbestand des Ernst-Haeckel-Archivs*, edited by Uwe Hoßfeld and Olaf Breidbach (Berlin: Verlag für Wissenschaft und Bildung, 2005); and *The Love Letters of Ernst Haeckel Written Between 1889–1903*, edited by Johannes Werner, translated by Ida Zeitlin (London: Methuen, 1930). Sources for this particular section include letters from Haeckel to his parents on November 27, 1852 (on poetry raising man out of the dust), and February 17, 1854 (calling his microscope "my wife"); the fragment of the poem from Allmers in *Haeckel und Allmers: Die Geschichte einer Freundschaft in briefen der Freunde*, edited by Rudolph Koop (Bremen: Arthur Geist, 1941); Haeckel's letter to Anna in which he speaks of "the Eldorado of Zoology" from October 21, 1859; Johannes Müller's monograph that inspired Haeckel and which appeared just after his death was titled *Über die Thalassicollen, Polycystinen und Acanthometren des Mittelmeeres* (1858); Heinrich Georg Brunn's German translation of Darwin's *On the Origin of Species* was titled *Über die Entstehung der Arten im Their-und Pflanzen-Reich durch natürlich Züchtung; oder, Erhaltung der vervollkommneten Rassen in Kampfe um's Daseyn* (Stuttgardt: Schweizerbart'sche Verlagshandlung, 1860). Haeckel delightedly reported to Darwin that Anna was calling him her "*German Darwin-man*" in a letter from August 10, 1864, existent in the Darwin Papers in Cambridge online. And Haeckel's quote on a "progressive metamorphosis" comes from his talk to the Society of German Natural Scientists and Physicians in Stettin, on the 17th of September 1863, titled "*Ueber die Entwickelungstheorie Darwins*" ("On Darwin's evolutionary theory"). Finally, Haeckel did really become the world expert on the Radiolaria: While he described 144 in his initial work on this group, later on he would be among a select group of renowned experts asked to examine the discoveries made by the HMS *Challenger*, a research vessel commissioned by the British government and the Royal Society of London to systematically explore the depths of all the world's oceans, from 1873 to 1876. Haeckel would go on to produce two thick folio volumes on the Radiolaria, plus a third with plates, reporting on no less than 4,318 species and 739 genera. See his monumental *Report on Radiolaria*, in the *Report on the Scientific Results of H.M.S. Challenger During the Years 1873–1876*, from 1887. It's phenomenal. Besides all these, Samuel Taylor Coleridge's poem *The Rime of the Ancient Mariner* from 1798 also provided inspiration.

6. "Cells and Embryos": Antonie van Leeuwenhoek's description of the contents of sperm comes, once again, from the reproduction of his letter to the Royal Society in F. J. Cole, *Early Theories of Sexual*

Generation (Oxford, 1930), and his correspondence with Harvey is from November 12, 1680, and can be viewed at De Digitale Bibliotheek voor Nederlandse Letteren 333. His prophetic words about "matter capable of being spun out"—which sounds a lot like DNA—are in the preface to *The Select Works of Antony van Leeuwenhoek*, edited and translated by Samuel Hoole (London: G. Sidney, 1800). The history of cell theory and of embryology became increasingly wound up with each other ever since Leeuwenhoek's death. They include, in Spallanzani's footsteps, the two French scientists, Jean Prevost (1790–1850) and Jean Duma (1800–1884), not mentioned in the text, who repeated Regnier de Graaf's experiment, showing (without ever seeing one) that eggs must be released from ovaries. Classical studies that I used for this chapter include Henry Harris, *The Birth of the Cell* (Yale, 1998), in which Harris translates primary sources for the first time into English as well as focusing on scientists outside the usual French, Italian, German, and English spheres who have received less attention, including Jan Purkinjê, Gabriel Gustav Valentin, and their collaborators in largely Polish Breslau whose work closely tracked that of Schwann and Schleiden; Joseph Needham, *A History of Embryology* (Cambridge, 1959), as well as a more recent book with the same title, edited by T. J. Horder, J. A. Witkowski, and C. C. Wylie (Cambridge, 1986), which describes embryology in the century after 1880; Paul Weindling and K. Figlio, who suggest that Virchow cribbed Remak's idea of *Omnis cellula e cellula* in their "Was Social Medicine Revolutionary? Rudolf Virchow and the Revolutions of 1848," *The Society for the Social History of Medicine Bulletin* 34 (1984): 10–18; and Lynn K. Nyhart, *Biology Takes Form: Animal Morphology and the German Universities 1800–1900* (Chicago, 1995), which tells the story of the central involvement of German scientists and their institutions. Laura Otis's *Müller's Lab* (Oxford, 2007) was once more useful for information on Remak, Schwann, and Virchow, and the more popular *The Song of the Cell*, by Siddhartha Mukherjee (Scribner, 2022) helped tie the strands together. A scholarly biography of Virchow is Irwin H. Ackerknecht's *Rudolf Virchow: Doctor, Statesman, Anthropologist* (University of Wisconsin Press, 1953), and those of you who like more popular history might enjoy Leslie Dunn's *Rudolf Virchow: Four Lives in One* (Leslie Dunn, 2016). In any case, do look at Virchow's classic *Cellular Pathology*, which was published in German in 1858 but is available in Franklin Chance's English translation from 1860. Besides including the quote on "the body as a cell state in which every cell is a citizen," it's a foundational book of our modern world. Also worthwhile, to get a sense for the history, is Johannes Müller's classic, and massive, *Elements of Physiology*, which was translated into English by William Baly in 1840, and Schleiden and Schwann's slightly earlier separate papers, "Contributions to our Knowledge of Phytogenesis," and "Microscopical Researches into the Accordance in the Structure and Growth of Animals and Plants," respectively. For those wanting to delve deeper into the history and philosophy of cell theory and of the science of development, see "Theories in Early Embryology," by Linda van Speybroeck, Dani de Waele, and Gertrudis van de Vijver, *Annual New York Academy of Sciences* 98 (2002): 7–49, which provides a good explanation of the epigenetic-preformationism debates, as well as Daniel J. Nicholson, "Biological Atomism and Cell Theory," *Studies in the History and Philosophy of Biological and Biomedical Sciences* 41 (2010): 202–211, which situates cell theory in an older tradition of biological atomism going back to Newton and Buffon, namely the belief that all living things must be understood by observing the indivisible constituent parts that compose them, which in themselves already exhibit all the characteristics proper to life. A succinct source on the significance of Karl Ernst von Baer is Arhat Abzhanov, "Von Baer's Law for the Ages: Lost and Found Principles of Developmental Evolution," *Trends in Genetics* 29, no. 12 (2013): 712–722. Besides providing the quote for the Meckel-Serres Law of Parallelism, it explains how von Baer's nineteenth-century ideas still resonate today. Finally, I am grateful to Janina Wellmann for discussions of metamorphosis, writ large. Her chapter "Metamorphosis in Images: Insect Transformation from the End of the Seventeenth to the Beginning of the Nineteenth Century," in the book *Drawing Processes of Life: Molecules, Cells, Organisms*, edited by Gemma Anderson-Tempini and John Dupré (Intellect,

2023), was especially helpful to me in making the leap from Maria Sibylla Merian to Haeckel.

7. "*Deus Sive Natura*": When Haeckel returned from Nice where his parents had sent him in the hope of overcoming the great sadness at the death of his wife, he returned with one thing on his mind. The letter to Darwin stating his intent is from July 7, 1864, and can be found online in the Darwin Correspondence Project. This was a philosophical time for Haeckel, in which he attempted to figure out where he stood on big questions. For a deeper dive into the Kantian philosophy that Haeckel was responding to, including the skepticism that there might be a "Newton of the blade of grass," see Robert J. Richard's *The Romantic Conception of Life: Science and Philosophy in the Age of Goethe* (Chicago, 2002). Though many of Haeckel's later popular books would spark controversy, his *Generelle Morphologie* from 1866 is a classic. Each of its chapters opened with a quote from Faust, including the one in this essay, also quoted in Richard's *The Tragic Sense of Life*. Still, many people, including both Darwin and even Darwin's intrepid "bulldog" T. H. Huxley, didn't like the book's combative tone. Darwin's letter to Haeckel to that effect is from August 18, 1866, and can be found in the Haeckel Correspondence in the Haeckel-Haus in Jena. *Generelle Morphologie* not only introduced the terms "ecology" and "phylum," and identified "protists" as a separate kingdom. It not only suggested that multicellular animals ("metazoans," another word Haeckel coined) arose from the aggregation of single-celled creatures; introduced countless new marine species; tackled the complex issue of biological individuality; and proposed a strong formulation of recapitulation in the form of the biogenetic law, but also introduced for the first time evolutionary trees as powerful tools for tracing lineages. The book on the evolution of languages by Haeckel's good friend at Jena, August Schleicher, which provided inspiration for this, is titled *Die Darwinische Theorie und die Sprachwissenschaft*, and was published, and dedicated to Haeckel, in 1863. For more on the biogenetic law, and its connection to the issue of individuality, see again Ruth G. Rinard, "The

Problem of the Organic Individual: Ernst Haeckel and the Development of the Biogenetic Law," *Journal of the History of Biology* 14, no. 2 (1981): 249–275. Haeckel's trip to the Canary Islands produced a beautiful book on Siphonophores titled *Zur Entwickelungsgeschichte der Siphonophoren* in 1869, which he submitted anonymously to a competition announced by the Utrecht Society for Arts and Sciences, and won a gold medal. Finally, I recommend a wonderful book by Rebecca Newberger Goldstein called *Betraying Spinoza: The Renegade Jew Who Gave Us Modernity* (Schocken, 2009) for those who want to get to know the man who coined the phrase *Deus sive Natura*.

8. "Axolotl": The tale of the Mexican axolotl's arrival and dispersal in Europe is told by Christian Reiß, Lennart Olsson, and Uwe Hoßfeld, in "The History of the Oldest Self-Sustaining Laboratory Animal: 150 Years of Axolotl Research," *Journal of Experimental Zoology B (Molecular Developmental Evolution)* 324B (2015): 393–404, to which I owe much of the information in this section. The axolotl's importance in evolutionary research is also featured in Stephen Jay Gould's *Ontogeny and Phylogeny*, which is the source of the Haeckel quote on the axolotl (Gould's essay "A Biological Homage to Mickey Mouse" is a wonderful primer on neoteny). But evolution isn't the only reason that axolotls are so interesting. Their amazing regenerative abilities are highlighted in Stéphane Roy and Samuel Gatien's review "Regeneration in Axolotls: A Model to Aim For!," *Experimental Gerontology* 43 (2008): 968–973, and made explicitly relevant to humans in a more recent article by W. A. Vieira, K. M. Wells, and C. D. McCusker, "Advancements to the Axolotl Model for Regeneration and Aging," *Gerontology* 66 (2020): 212–222. It's hard to pinpoint just where humans begin to age, since genetic and environmental factors combine in complex ways to affect the process. And clearly, neoteny cannot fully explain regeneration: Like the axolotl, we too are largely neotenic, but no young or old human to date has successfully grown back a hand, much less a brain. Still, those of you interested in the work being done to unmask the genetics involved should read Juan Caballero-Perez et al., "Transcriptional

Landscapes of Axolotl (*Ambystoma mexicanum*)," *Developmental Biology* 433 (2018): 227–239; and S. Nowoshilow et al., "The Axolotl Genome and the Evolution of Key Tissue Formation Regulators," *Nature* 554 (2018): 50–55. While it was the Italian Catholic priest Lazzaro Spallanzani who first discovered their unsurpassed ability to regenerate entire body parts more than two hundred years ago, the twentieth-century British embryologist and developmental biologist Conrad Hal Waddington was one of the first to appreciate the cancer resistance of axolotls, in his seminal paper "Cancer and the Theory of Organisers," *Nature* 135 (1935): 606–608. Another British developmental biologist, Gavin De Beer, clarified in the 1930s that while Haeckel defined heterochrony as a change in the timing of the embryonic development of one organ relative to the rest in the same animal, it should actually be a change relative to the same organ in the animal's predecessor; when we follow this definition, rather than Haeckel's, it turns out that humans really are generally neotenic relative to apes. Where we stand today on why and how we age is succinctly summarized in K. Jin, "Modern Biological Theories of Aging," *Aging and Disease* 1, no. 2 (2010): 72–74, and Andrew Steele provides a fascinating tour through the science of stunting our aging in *Ageless: The New Science of Getting Older Without Getting Old* (Doubleday, 2021). For an eye-opening forecast of where the world's population is going, see Professor Stein Emil Vollset et al., "Fertility, Mortality, Migration and Population Scenarios for 195 Countries and Territories from 2017 to 2100: A Forecasting Analysis for the Global Burden of Disease Study," *The Lancet* 395, no. 10258 (2020): 1285–1306.

There are a number of first-rate biographies of J. M. Barrie, the creator of Peter Pan. If you have to choose one book, I recommend Andrew Birkin's *J. M. Barrie and the Lost Boys* from Yale University Press, originally published in 1979 and reissued with a new preface in 2003 to mark Peter Pan's centennial. A more recent academic book is *Peter Pan and the Mind of J. M. Barrie* (Cambridge Scholars Publishing, 2016) by the neuroscientist Rosalind Ridley. Looking carefully at Barrie's further books, *Peter Pan in Kensington Gardens* (1906) and *Peter and Wendy* (1911), Ridley argues that, informed by a post-Darwinian perspective on the evolutionary origins of human behavior, Barrie described the limited mental abilities of infants and animals in order to illuminate the structure of human adult cognition. From time to time the popular press latches on to the accusation that Barrie was a pedophile: See Justin Picardie's July 13, 2008, article in the *Telegraph*, "How Bad Was J. M. Barrie?" Watching the 2004 film *Finding Neverland* (based on Allan Knee's play), with Johnny Depp as Barrie and Kate Winslet as Sylvia Llewelyn Davies, and listening to Kate Bush's haunting song "In Search of Peter Pan" from her 1978 album *Lionheart*, I prefer to believe the denial of Nico, the youngest of the Llewelyn Davies boys, who swore that Barrie had never behaved inappropriately. "I don't believe that Uncle Jim ever experienced what one might call 'a stirring in the undergrowth' for anyone—man, woman, or child," he is on record as stating. "He was an innocent—which is why he could write Peter Pan." Poor Nico lost his older brothers George, to World War I, and Michael, to a river. Tragically, some years after Barrie's death, Peter himself found his death by throwing himself in front of a moving train. This section greatly benefited from a fantastic exhibition at the Israel Museum, named *Peter and Pan: From Ancient Greece to Neverland*, curated by Rachel Caine Kreinin, David Mevorah, and Morag Wilhelm, and designed by Michal Aldor—which ran from June to December 2019. Peter embodies contrasts, like the Greek god Pan, son of Hermes, part-goat, part-man: brave and fearless, reckless, and in every way ambiguous. Sexless and virile, altruistic and avaricious, Pan represents human hopes, but also everything that we fear. Barrie masked his roots, abducting him from his ancient Peloponnesian cave and dropping him in Neverland. Satyrs and nymphs were replaced by Captain Hook and Tinker Bell. Still, through his modern spectacles, Barrie succeeded in catching a glimpse of what this all meant: the wonder of never growing old.

Finally, the role of the axolotl in felling Haeckel's biogenetic law is fascinating. Walter Garstang (1868–1949) was a marine zoologist at the University of Leeds who was taken by the axolotl. His "The Theory of Recapitulation: A Critical Re-statement of the Biogenetic Law,"

Zoological Journal of the Linnean Society 35 (1922): 81–101, is a historical document, and Maurizio Esposito has a nice recent article on the subject, "Beyond Haeckel's Law: Walter Garstang and the Evolutionary Biology That Might Have Been," *Journal of the History of Biology* 53 (2020): 249–268. Looking at the axolotl, Garstang was moved to write a poem in his posthumous book *Larval Forms with Other Zoological Verses* (Basil Blackwell, Oxford, 1951). It goes like this:

Ambystoma's a giant newt who rears in swampy waters,
as other newts are wont to do, a lot of fishy daughters:
These Axolotls, having gills, pursue a life aquatic,
But, when they should transform to newts, are naughty and erratic.

They change upon compulsion, if the water grows too foul,
for then they have to use their lungs, and go ashore to prowl:
but when a lake's attractive, nicely aired, and full of food,
they cling to youth perpetual, and rear a tadpole brood.

And newts Perennibranchiate have gone from bad to worse:
They think aquatic life is bliss, terrestrial a curse.
They do not even contemplate a change to suit the weather,
But live as tadpoles, breed as tadpoles, tadpoles all together!

9. "Riddle": Haeckel's views on man have been a controversial subject. Here, I wished as loyally as possible to present his voice on the matter. Unquestionably, one of Haeckel's contributions to evolutionary studies was his stem-trees, and here his friend Schleicher's influence was great. See again August Schleicher's *Die Darwinische Theorie und die Sprachwissenschaft* from 1863, where he develops the tool of the stem-tree to indicate distance and closeness between languages. When Haeckel introduced stem-trees in his popular book *Natürlische Schöpfungsgeschichte* in 1868, there was a surge in popular literature in Germany about Native Americans, who were depicted romantically as ideals of the noble savage. This probably accounted for their high standing in Haeckel's hierarchy, though they were demoted in the eighth edition in 1889, replaced by the Indians from the subcontinent. Like other naturalists of his or any day, Haeckel was strongly influenced by cultural biases, but his more exacting anatomy led to the prescient description of heterochrony, about whose evolutionary role more can be learned briefly in Stephen Jay Gould, "On the Importance of Heterochrony for Evolutionary Biology," *Systematic Zoology* 28, no. 2 (1979): 224–226, and in more depth in Gould's book *Ontogeny and Phylogeny* (Harvard, 1977). Gould was very disparaging of Haeckel regarding his views on different races, as well as his so-called faked developmental images—more about that in the next section—but recognized the importance of his introduction of the term *heterochrony*. Important, too, was Haeckel's suggestion that a "missing link" be searched for, a task taken on by Eugène Dubois, whose grateful letter to Haeckel from December 24, 1895, after discovering Java Man can be found in the Haeckel Correspondence at Haeckel-Haus, in Jena. After the popular success of *The Natural History of Creation*, Haeckel came back in 1874 with *Anthropogenie; oder, Entwickslungsgeschichte des Menchen* (translated to English as *The Evolution of Man*), which was printed in six editions, and in 1899 with his most famous offering, *Die Welträtsel*, which literally means "the world puzzles," but was titled *The Riddle of the World* in English. The popular success Haeckel was gaining drew much fire, from both the church and many scientific colleagues. T. H. Huxley's supportive letter quoted in the text is from December 28, 1974; *The New York Times* article "Haeckel Kills the Soul" ran on May 8, 1905; and the Hampstead Congregationalist Church preacher's warning is quoted in R. F. Horton, "Ernst Haeckel's 'Riddle of the Universe,'" *Christian World Pulpit* 63 (June 10, 1903): 353–356. Haeckel was inspired to write *The Riddle of the Universe* as a reply to his fellow German Darwinist Emil du Bois-Reymond's philosophical view that there are certain problems, like consciousness, that man will never solve. Haeckel doubtless thought of himself as more scientific, or what we would call more hard-headedly positive, but Bois-Reymond doubtless thought the opposite: Haeckel's talk of the "soul" of cells and "life" of crystals seemed to him teleological, and unscientific. For a modern appreciation in this vein, see Gabriel Finkelstein, "Haeckel and du Bois-Reymond: Rival German Darwinists," *Theory in Biosciences* 138 (2019): 105–112, and for Haeckel's perspective—anything written by his biographer, Robert J. Richards. Those of you who feel the need to return to Nietzsche's Superman theory (what he called the *Übermensch*), it's in his *Thus Spoke Zarathustra*, published in four volumes between 1883 and 1885. And the history of Haeckel's Phyletic Museum was beautifully presented in a book sponsored by the museum on the occasion of its centennial in 2008 by Martin S. Fischer, Gunnar Brehm, and Uwe Hoßfeld, titled *Das Phyletische Museum in Jena*.

10. "Lucy": Scientists define metamorphosis as either "radical" or "dramatic" post-embryonic development, but that's somewhat arbitrary. After all, who gets to decide what "dramatic" or "radical" really means? To view puberty in humans as a form of metamorphosis is to acknowledge that the biological definition of metamorphosis exists on a continuum: We may not turn from caterpillars into butterflies, but we sure do go through a lot of post-embryonic change. To learn more about the considerable physical changes that happen during human adolescence, featuring brain-driven interactions between hormones and the nervous system, see Cheryl L. Sisk and Douglas L. Foster, "The Neural Basis of Puberty and Adolescence," *Nature Neuroscience* 7 (2004): 1040–1047. Adolescence is so peculiar that legal minds have recently begun to argue that the usual binary child/adult should be abandoned in favor of a tri-partite classification. See Elizabeth S. Scott and Laurence Steinberg's book *Rethinking Juvenile Justice* (Harvard University Press, 2010), in which they make the case for creating a third legal category: the adolescent. A further, more recent source on the science involved is Jean-Pierre Bourguignon, Jean-Claude Carel, and Yves Christen, eds., *Brain Crosstalk in Puberty and Adolescence* (Springer, 2015), but consider the limits of science in this debate, as set out in my essay "Unformed Minds: Juveniles, Neuroscience and the Law," *Studies in the History and Philosophy of the Biological and Biomedical Sciences* 44, no. 3 (2013): 455–459. On the role of sexual selection in the evolution of females and males—their relationships, behaviors, and physiologies—see Richard R. Prum, *The Evolution of Beauty: How Darwin's Forgotten Theory of Mate Choice Shapes the Animal World—and Us* (Anchor, 2018), and my review of it (upon which I draw in this section), "Darwin's *Really* Dangerous Idea," *Science and Education* 28, no. 6 (2019): 803–812. For an entertaining, if speculative and rather too reductive, account of what the period of adolescence may actually be for, from an evolutionary perspective, read David Brainbridge's 2009 book *Teenagers: A Natural History.* A general and excellent primer on the relationship of our human natures to different apes is Frans De Waal, *Our Inner Ape: A Leading Primatologist Explains Why We Are Who We Are* (New York: Riverhead Books, 2006), and two solid treatments of our moral natures from an evolutionary perspective are Richard Joyce, *The Evolution of Morality* (MIT Press, 2006); and *Moral Origins: The Evolution of Virtue, Altruism and Shame* (Basic Books, 2012), by Christopher Boehm. Boehm also has an excellent book titled *Hierarchy in the Forest: The Evolution of Egalitarian Behavior* (Harvard University Press, 2001), which relates to the subject of this section. As for a new assessment of the sex lives of primates, see Esther Clarke et al., "Primate Sex and Its Role in Pleasure, Dominance and Communication," *Animals* 12, no. 23 (2022): 3301. Finally, when it comes to Lucy, read the 1981 account by the discoverer himself, Donald Johansen, with Maitland Edey, *Lucy: The Beginnings of Humankind*, as well as a sequel from 2009, this time with Kate Wong, *Lucy's Legacy: The Quest for Human Origins.* They'll get you thinking about just how wonderfully serendipitous human evolution is.

11. "Strauss": There are reasons why Haeckel remains a dark figure of history in many minds: his combative personality, the charges of fraud, the adoption of some of his ideas by the Nazis. His biographer Richards does his best to exonerate him of the harsher charges, agreeing that some of the images and woodcuts he used were a serious lapse in judgment, but pointing out that they were made in a popular book, not a scientific treatise, that woodcuts were expensive to produce, and since the best science of the time agreed that the morphological structures of early embryos of different species are virtually indistinguishable, it made economic sense to simply replicate the same images as a device to bring home the message to the public. I believe that while much of the apology for Haeckel's fudging of images remains unconvincing, Richards successfully debunks the myth that Haeckel provided the intellectual foundations for Nazism. But the best way to develop an opinion on these matters is to check out the facts for yourselves: On the charges of fraud, compare Richards's "Haeckel's Embryos: Fraud Not Proven," *Biology and Philosophy* 24 (2009): 147–154; and E. Watts, G. S. Levit, and U. Hoßfeld, "Ernst Haeckel's Contribution to Evo-Devo and Scientific Debate: A Re-evaluation of Haeckel's Controversial Illustrations in U. S. Textbooks in Response to Creationist Accusations," *Theory in Biosciences* 138 (2019): 9–29; to Richardson's 1997 *Science* article "Haeckel's Embryos: Fraud Rediscovered," and his more considered piece (again with Gerhard Keuck), "A Question of Intent: When Is a 'Schematic' Illustration a Fraud?," *Nature* 410 (2001): 144. Finally, have a look at Nick Hopwood's magisterial study of the controversy surrounding *Haeckel's Embryos: Images Evolution, and Fraud* (Chicago, 2015), in which he tries to explain why the copying of images, the epitome of the unoriginal, allows images to become creative, contested, and of consequence. If you read German, you might also want to look at Wilhelm His's nineteenth-century attack on Haeckel in *Unsere Körperform und das physiologische Problem ihrer Entstehung: Briefe an einen befreundeten Naturforscher* (Leipzig, 1874–1875). On Haeckel's ties to Nazism, compare Richards's *The Tragic Sense of Life* (University of Chicago Press, 2008) to Daniel Gasman, *The Scientific Origins of National Socialism: Social Darwinism in Ernst Haeckel and the German Monist League* (Science History Publications, 1971), and Stephen Jay Gould, *Ontogeny and Phylogeny* (Harvard, 1977), from which the quote in the text comes. Compare, too, Peter Weikart's *From Darwin to Hitler: Evolutionary Ethics, Eugenics, and Racism in Germany* (Palgrave

Macmillan, 2004) to Richards's own *Was Hitler a Darwinian?* (Chicago, 2013), which takes a very different view. When both were alive, Haeckel and Darwin were on nothing but good terms, each respecting the other as a scientist and an ally. In "The Break Up Between Darwin and Haeckel," *Theory in Biosciences* 138 (2019): 113–117, Nicolaas Rupke takes a look at the social and political, rather than scientific, issues that brought about a separation between the two in many people's minds. Some historians continue to argue that Darwin and Haeckel held completely different approaches to evolution: Peter Bowler, for instance, holds that while Darwin eschewed all forms of teleology, including rejecting the biogenetic law, Haeckel should be viewed as a late representative of romantic German *Naturphilosophie*, and not a modern scientist, much less a true representative of Darwin. See his book *The Non-Darwinian Revolution* (Johns Hopkins University Press, 1988), and the exchange in "A Bridge Too Far," *Biology and Philosophy* 8 (1993): 98–102. That Haeckel was adored by many in his day is uncontroversial; the Isadora Duncan scene, for example, including her letter to him from May 8, 1910, is described in Richards's biography. Described in Richards, too, are Haeckel's pacifism before, and patriotism during, the war; his spat with Plate; his affair with Frida; and his death. It is impossible to know whether, like Strauss, Haeckel would have tried to walk in between the drops in order to go on doing his science, but I think Richards is correct in judging that he would have felt repugnance toward the Nazis. As for the question of Haeckel's overall importance in the history of biology, Georgy S. Levit and Ute Hoßfeld provide a balanced analysis in "Ernst Haeckel in the History of Biology," *Current Biology* 29 (2019): R1276–R1284, from which the quote on the triumph of Darwinism being "unthinkable without Haeckel" comes. For the continued relevance of Haeckel's science, in particular his biogenetic law, see Michael K. Richardson and Gerhard Keuck's "Haeckel's ABC of Evolution and Development," *Biological Review* 77 (2002): 495–528; L. Olsson, G. S. Levit, and U. Hoßfeld, "The 'Biogenetic Law' in Zoology: From Haeckel's Formulation to Current Approaches," *Theory in Biosciences* 136 (2017): 19–29; and M. Usesaka, S. Kuratani, and N. Irie, "The Developmental Hourglass Model and Recapitulation: An Attempt to Integrate the Two Models," *Journal of Experimental Zoology B* 338 (2022): 76–86. Richardson's "Theories, Laws and Models in Evo-Devo" in that same issue (pp. 36–61) is also clarifying.

There are a number of good sources on Strauss, including Bryan Gilliam's *The Life of Richard Strauss* (Cambridge University Press, 1999), and Charles Youmans's more recent *Richard Strauss's Orchestral Music and the German Intellectual Tradition: The Philosophical Roots of Musical Modernism* (Indiana University Press, 2005). For a heartbreaking description of American soldiers arriving at Strauss's villa in Garmisch toward the end of World War II, see the relevant chapter in Alex Ross's book *The Rest Is Noise: Listening to the Twentieth Century* (Picador, 2008). I am most grateful to Jeremy Eichler for sharing with me his hauntingly beautiful manuscript, "The Truth of Music: Four Composers and the Memory of the Second World War," before it was published as *Time's Echo: The Second World War, the Holocaust, and the Music of Remembrance* (Knopf, 2023). It includes an insightful description of Strauss and his Jewish librettist Stefan Zweig's fraught relationship, as well as a deep interpretation of *Metamorphosen*, from which the "I made music under the Kaiser," "infinite reaches of heaven," "death mask," "opacity of self," and Nietzsche quotes come. Of the musical composition, Eichler writes: "Gone are the glittering facades of irony and wit. Gone are the liberated heroes of the early tone poems. Gone is the proudly modern pose of objectivity, the lofty sense of authorial detachment from his own music. In its place is an almost disorienting sense of sincerity." For more on the historical appraisal of Strauss and his relationship to the Nazis, see Pamela M. Potter's article, "Strauss and the National Socialists: The Debate and Its Relevance," in Bryan Gilliam's edited volume *Richard Strauss: New Perspectives on the Composer and His Music* (Duke University Press, 1992), as well as a disparaging take by Oliver Hilmes in his bestselling book *Berlin 1936: Sixteen Days in August* (The Bodley Head, 2018). The "hunger for wholeness" quote comes from Anne Harrington, in her book *Re-enchanted Science: Holism in German Culture from Wilhelm II to Hitler* (Princeton University Press, 1996). And the final Goethe quote comes from his poems *Faust*, *Prometheus*, and *God and the World*, as quoted by Haeckel in the final lines of his *The Riddle of the Universe*.

12. "Eel": The European eel, *Anguila anguila*, is a force of nature, and has a storied history besides. Patrik Svensson marries a beautiful personal journey with the eel's story in *The Book of Eels: Our Enduring Fascination with the Most Mysterious Creature in the Natural World* (Ecco, 2020), which provided this chapter both facts and inspiration. So, too, did Lucy Cooke's delightful *The Unexpected Truth About Animals* (Doubleday, 2016), which is the source of the quote on Freud. For those of you interested in diving deeper into the nature of eel metamorphosis, see the suggestion that silvering is a kind of pubertal, rather than a classic, metamorphic event, in S. Aroua, M. Schmitz, S. Baloche, B. Vidal, K. Rousseau, and S. Dafour, "Endocrine Evidence That Silvering, a Secondary Metamorphosis in the Eel, Is a Pubertal Rather Than a Metamorphic Event," *Neuroendocrinology* 82 (2005): 221–232. Recently, researchers have suggested that glass eels use a tidal phase–dependent magnetic compass for orientation to reach the European coasts; see Alessandro Cresci et al., "Glass Eels (*Anguila anguila*) Imprint the Magnetic Direction of Tidal Currents from their Juvenile Estuaries," *Communications Biology* 2, no. 366 (2019). But how precisely European eels migrate back to the Sargasso Sea, and how they mate, remains a mystery. Cooke quotes the German marine biologist Leopold Jacoby, expressing his frustration: "It is certainly somewhat humiliating to men of science, that a fish which is commoner in many parts of the world than any other fish . . . which is daily seen at the market and on the table, has been able in spite of the powerful aid of modern science, to shroud the manner of its propagation, its birth, and its death in darkness, which even to the present day has not been dispelled." Jacoby wrote these words in 1879, and still no eel has been caught being made. The eel remains just as slippery as ever.

Part Three: "What Is the Self?"

1. "Mount Monadnock": My first conversation with Lynn Riddiford and Jim Truman took place over Zoom on May 29, 2023. Jim did most of the talking, and Lynn most of the smiling. Both were tickled by the fact that they were going to feature in the same book as Maria Sibylla Merian and Ernst Haeckel. Merian was a heroine of theirs. They even owned a 1705 uncolored folio of her Surinam book, which they had bought near Oxford, the "pièce de résistance" of their library, they said. Ever since, I have enjoyed a rigorous and candid conversation with them over email, and a wonderful visit together on San Juan Island, where they generously invited me into their home, as well as their lab at Friday Harbor. Over time, Jim and Lynn and I developed a back-and-forth that allowed me to get to know them, and I am deeply thankful. Besides the personal contact, there were written sources that helped me: all of their scientific papers, hundreds in number and still counting, but also a few interviews and profiles here and there, including: Marline E. Rice, "Lynn M. Riddiford: From the Barnyard to Harvard Yard," *American Entomologist* 65, no. 2 (2019): 80–86; Sujata Gupta, "QnAs with Lynn M. Riddiford," *Proceedings of the National Academy of Sciences* 110, no. 31 (2013): 12501–12502; Lynn's own "A Life's Journey Through Insect Metamorphosis," *Annual Reviews of Entomology* 65 (2020): 1–16; and Rivka Galchen's article about Jim's work, "What Insects Go Through Is Even Weirder Than We Thought," *The New Yorker*, August 31, 2023. *Monadnock* is the title of a famous poem by Ralph Waldo Emerson, which he began writing at the top of the mountain on the morning of May 3, 1845, and in which he calls the mountain the "constant giver," and the Henry David Thoreau quote comes from his poem *With Frontier Strength Ye Stand Your Guard*, from 1843. Thoreau climbed the mountain in 1844, 1852, 1858, and 1860 and recorded extensive botanical and geological observations in his journal (there's even a bog on the mountain named after him). "Almost without interruption we had the mountain in sight before us," he wrote on his first ascent, "its sublime gray mass—that antique, brownish-gray, Ararat color. Probably these crests of the earth are for the most part of one color in all lands, that gray color of antiquity, which nature loves." For more on this topic, see Elliott Allison's essay "Thoreau of Monadnock," *Thoreau Journal Quarterly*, October 1973.

2. "Rumi": I first learned of Sufism and of Rumi when I visited a mosque in Istanbul as a youth to see the dance of the whirling dervishes. It was Rumi's son, continuing his music-loving father's legacy, who founded this group of mystical dancers, also known as the Mevlevi Order, famous for the so-called Sema ceremony in which, wearing conical felt hats and white skirts unfurling at their base like flowers, lost in mystical meditation, they twirl and twirl and twirl. Watching them is a transporting experience. As our family moved together into the third trimester, Rumi's poems suffused our household. So did music inspired by him, especially this soothing track, composed by Levon Minassian, starting with the song "Siretzi Yaras Daran" ("They Have Taken the One I Love"): https://www.youtube.com/watch?v=5I5otf98JKY. To learn more about Rumi, read his biographer Franklin Lewis's *Rumi: Past and Present, East and West* (Oneworld Publications, 2008), and to encounter him directly, see *The Essential Rumi: New Expanded Edition*, translated by Coleman Barks (HarperOne, 2004), from which the quotes and the poem come.

The great American anthropologist Clifford Geertz reminds us that at least some conception of what a human is, as opposed to a rock or a butterfly, is universal. But in the West, we have come to think of the "self," as he writes, as a "bounded, unique, more or less integrated cognitive and motivational universe; a dynamic center of awareness, motion, judgment, and action organized into a distinctive whole and set contrastively against both other such wholes and against a social and natural background." And other traditions, particularly non-Western ones like the ones Geertz himself studied, have entirely different understandings of personhood. It's important to remember that when considering the ways in which scientific and philosophical and cultural thought on metamorphosis have impacted one another in our tradition. For more on how selfhood has been imagined differently in the Javan, Balinese, and Moroccan contexts, and on the impossibility of really getting into the mind of a person from another culture, see Geertz's "'From the Native's Point of View': On the Nature of Anthropological Understanding," *Bulletin of the American Academy of Arts and Sciences* 26, no. 1 (1974): 25–45.

3. "Sea Squirt, Starfish": To learn more about the amazing ascidians, which are a class of tunicates, have a look at this website from the University of Sorbonne, in France: https://digital-marine.sorbonne-universite.fr/index.php/tunicates. But watch out: You may very well fall in love with these distant relatives. I think Darwin did, pegging them in 1871 as our ancient ancestors in his *Descent of Man*. David Starr Jordan, on the other hand, felt they had morally digressed: See his paper, with Edwin Grant Conklin, Frank Mace Macfarland, and James Perrin Smith, "Foot-Notes to Evolution: A Series of Popular Addresses on the Evolution of Life" (D. Appleton, 1898). For the historians among you, the original Alexander Kovalevsky paper, in German, in which he figured out the sea squirt life cycle, is titled "Entwickelungsgeschichte der einfachen Ascidien," in *Mémoirs De L'académie Impériale Des Sciences De St. Pétersbourg*, VII Série, Tome X N° 15, 1866. And for the scientists—to delve deeper into the details we now know—read Noriyuko Satoh, "A Deep Dive into the Development of Sea Squirts," *Nature* 571 (2019): 333–334. Those of you who are interested in how minds and brains first evolved, look at *The Ancient Origins of Consciousness: How the Brain Created Experience*, by Todd E. Feinberg and John M. Mallatt (MIT Press, 2016), and the challenging and brilliant book by Eva Jablonka and Simona Ginsburg *The Evolution of the Sensitive Soul: Learning and the Origins of Consciousness* (MIT Press, 2019). Neither may unravel the mystery of what it feels like to be a sea squirt, but they're sure to get you thinking.

An early and quaint description of the unusual development of the starfish can be read in the Scottish physician James F. Gemmill's 1915 article "Twin Gastrulae and Bipennariae of *Luidia sarsi*, Düben and Koren," *Journal of the Marine Biological Association of the United Kingdom* 10, no. 4 (1915): 577–589, in which the author writes "It soon became evident that the teratological type in question, namely double monstrosity, was about to receive a more varied expression, and to attain a more advanced stage in development, than it had ever before been my good fortune to find in any starfish culture." Gemmill, a fellow of the Royal Society who shared a strong affinity to the Scottish poet Robert Burns, washed up downstream of the Tay Rail Bridge in 1926, having apparently committed suicide, unmarried and childless, following a bout of depression. The journal in which he first published his findings continued to be a platform for studies of the unusual starfish: See Douglas P. Wilson's 1978 article there, titled "Some Observations on Bipinnariae and Juveniles of the Starfish Genus *Luidia*," *Journal of the Marine Biological Association of the United Kingdom* 58, no. 2 (1978): 467–478. For an evo-devo approach, see B. J. Sly, M. S. Snoke, and R. A. Raff, "Who Came First—Larvae or Adults? Origins of Bilaterian Metazoan Larvae," *International Journal of Developmental Biology* 47 (2003): 623–632; we'll have more to say about this question in the section on Donald I. Williamson. Meanwhile, a recent paper explains for the first time how starfish (and other echinoderms like sea urchins) developed their fivefold symmetry, unusual among bilaterians. They show that the five-armed starfish is really a kind of disembodied head walking about the sea floor on its lips: See L. Formerly et al., "Molecular Evidence of Anteroposterior Patterning in Echinoderms," *Nature* 623 (2023): 555–561. To further whet your own lips, a short film on the stunning *Luidia sarsii* metamorphosis can be found on YouTube: https://www.youtube.com/watch?v=dbyR6tmkKV4.

4. "Evo-Devo": The history of genetics and the history of embryology have been largely written apart from one another, mirroring the divorce of the two for so much of the twentieth century in biology itself. On the genetics side, a first-class history is James Schwartz's *In Pursuit of the Gene: From Darwin to DNA* (Harvard University Press, 2010), as is Robert E. Kohler's *Lords of the Fly: Drosophila Genetics and the Experimental Life* (University of Chicago Press, 1994), which focuses on the fruit fly Drosophila, the workhorse of twentieth-century genetic research. The "father" of genetics, Gregor

Mendel, has been receiving newfound attention in recent years: See Gregory Radick's *Disputed Inheritance: The Battle over Mendel and the Future of Biology* (University of Chicago Press, 2023), for a deeply researched and novel interpretation of what Mendel did, and how what he did was understood by those who "rediscovered" him. A good summary and corrective to the received view is Kostas Kampourakis, *How We Get Mendel Wrong and Why It Matters* (Routledge, 2024). On the evolutionary synthesis, Ernst Mayr and William Provine, eds., *The Evolutionary Synthesis: Perspectives on the Unification of Biology* (Harvard University Press, 1998); and Betty Smocovitis, *Unifying Biology: The Evolutionary Synthesis and Evolutionary Biology* (Princeton University Press, 1996), remain the best sources. As for the history of embryology and its heroes after 1880, *A History of Embryology*, edited by T. J. Horder, J. A. Witkowski, and C. C. Wylie (Cambridge, 1986), provides a host of perspectives. Garland Allen has a nice essay about Driesch flipping from strict materialist to vitalist called "Rebel with Two Causes: Hans Driesch," in *Rebels, Mavericks and Heretics in Biology*, edited by Oren Harman and Michael Dietrich (Harvard University Press, 2008). And the story of Hans Spemann and Hilde Mangold is told beautifully in Spemann's autobiography, *Hans Spemann: Furschung Und Leben* (Severus Verlag, 2012), some of whose more philosophical materials are translated by Victor Hamburger in *Journal of the History of Biology* 32, no. 2 (1999): 231–243. Hilde Mangold was amazing: Performing microtransplants in newts without the use of antibiotics while preventing infection showed just how talented she was. An English translation of the title of her dissertation would be "Induction of Embryonic Primordia by Implantation of Organizers from a Different Species," and Spemann and Mangold's original paper on the organizer in German is "über Induktion von Embryonalanlagen durch Implantation artfremder Organisatoren," *Archiv für Mikroskopische Anatomie und Entwicklungsmechanik* 100, nos. 3–4 (1924): 599–638. Much has been written about Spemann's philosophy. On the political side, it hasn't escaped the notice of a number of scholars that the top-down, controlling organizer was a metaphor Spemann knew would play out well in a German culture longing for a powerful state following the turmoil of the Great War; they mention the fact that he seems to have given a *Sieg Heil* at the conclusion of his Nobel speech, and that Otto Mangold was a proud Nazi. But it remains unclear how much of this was just performance. On the philosophy of science side, contrast Tim Horder and Paul Weindling's "Hans Spemann and the Organiser," in *A History of Embryology* (pp. 183–242), which sees him thinking as a throwback to nineteenth-century *Naturphilosophie*, and a more recent appraisal by Christina Brandt, "Vitalism, Holism, and Metaphorical Dynamics of Hans Spemann's 'Organizer' in the Interwar Period," *Journal of the History of Biology* 55 (2022): 285–320, which tries to paint Spemann's philosophy of science as more modern and rooted in the biology of his day. Quotes from Spemann's lecture in America titled "Embryonic Induction" come from a reprint in *American Zoologist* 27 (1987): 575–579. Haeckel's letter to Darwin on siphonophores is from February 1868, and can be found online at the Darwin Correspondence Project, as can Darwin's letter to Asa Gray from September 10, 1860, lamenting that no one is paying attention to the importance of embryology. Darwin's own mistaken theory of inheritance appeared in his book *Variation of Animals and Plants Under Domestication* in 1868. Finally, as we shall see ahead, development and evolution are being brought together again. An early source providing historical and philosophical context for this old-new marriage of disciplines is Manfred D. Laubichler and Jane Maienschein's edited volume *From Embryology to Evo-Devo: A History of Developmental Biology* (MIT Press, 2007).

5. "Wigglesworth": The *New Yorker* poem by John Updike published on March 25, 1955, probably didn't thrill Vincent Wigglesworth. The young Updike, just twenty-two years old, had been taken by the funny name of the British FRS, and his funny title—recently appointed the Quick Professor of Zoology back at Cambridge. In the wake of World War II and the bomb, science had gained new status. Rather than a comment on a man he'd never met, "V. B. Nimble, V. B. Quick" was a humanist's dig at scientists in general, with their self-promotion and unreliable time-keeping and the quickness and detachment in which they move from one project to the next.

V. B. Nimble, V. B. Quick

V. B. Wigglesworth wakes at noon,
Washes, shaves and very soon
Is at the lab; he reads his mail,
Swings a tadpole by the tail,
Undoes his coat, removes his hat,

Dips a spider in a vat
Of alkaline, phones the press,
Tells them he is F.R.S.,
Subdivides six protocells,
Kills a rat by ringing bells,

Writes a treatise, edits two
Symposia on "Will man do?,"
Gives a lecture, audits three,
*Has the ***** club in for tea,*
Pensions off an ageing spore,

Cracks a test tube, takes some pure
Science and applies it, finds,
His hat, adjusts it, pulls the blinds,
Instructs the jellyfish to spawn,
And, by one o'clock, is gone.

Condensing the "quickness" of a lifetime into a single sentence, twenty cadences long, Updike showed talent but probably chose the wrong subject to embarrass. Wigglesworth was a consummate loner, almost a recluse, publishing 264 papers of which only 19 were with collaborators, and living in his college quarters all week rather than lose focus traveling back home to his family. He was hardly a brash mover-and-shaker, and his interest lay far away from the limelight.

Extremely helpful to me in telling Wigglesworth's story was Frank Ryan's excellent book *The Mystery of Metamorphosis: A Scientific Detective Story* (Chelsea Green Publishing, 2011), especially pages 85–144. For those who would like to delve more deeply into Wigglesworth and his life, there are a number of other good sources. To begin with, Wigglesworth's own important books *The Principles of Insect Physiology* (1939) and *The Physiology of Insect Metamorphosis* (1954); as well as his classic papers "Factors in Controlling Moulting and 'Metamorphosis' in an Insect," *Nature* 131 (1934): 725–726; and "The Physiology of Ecdysis in Rhodinus prolixus (Hemiptera). II. Factors Controlling Moulting and 'Metamorphosis,'" *Quarterly Journal of Microscopical Science* 77 (1934): 191–222. Biographical details can be found in "Some Memories: Interview with Sir Vincent Wigglesworth, Fellow and Emeritus Quick Professor of Biology," *The Caian* (1979): 30–44; and in Michael Locke, "Sir Vincent Brian Wigglesworth, C. B. E., 17 April 1899–12 February 1994," *Biographical Memoirs of Fellows of the Royal Society* 42 (1996): 540–553, where his kids' half-joking lament comes from. Wigglesworth's own ideas about applied versus pure research are offered in his "Experimental Biology, Pure and Applied," *Journal of Experimental Biology* 55 (1971): 1–12, and the entirety of his papers are usefully available online at the Churchill Archive Centre at the University of Cambridge. Carroll Williams's historical paper on juvenile hormone is "The Juvenile Hormone of Insects," *Nature* 178 (1956): 212–213. More on him can be found in: A. M. Pappenheimer Jr., "Carroll Milton Williams: December 2, 1916–October 11, 1991," *Biographical Memoirs*, vol. 68, National Academy of Science Press, (1995), 413–433. (It's important to note for historical accuracy that Wigglesworth and Williams weren't alone in elucidating the hormonal control of metamorphosis; see, for example, Jean-Jacques Bounhiol, "Recherches expérimentales sur le déterminisme de la métamorphose chez les lépidoptères," *Bulletin Biologique de la France et de la Belgique* 24 (1938): 1–199; and S. Fukuda, "The

Hormonal Mechanism of Larval Molting and Metamorphosis in the Silkworm," *Journal of the Faculty of Science Tokyo University Section IV* 6 (1944): 477–532.) Though Wigglesworth lay the ground for insect physiology, in a sense he was superseded by Williams, who spent the year 1955 as a visitor in his lab. Williams was a brilliant experimenter, like the Englishman, inventing new methods, such as using carbon dioxide as an anesthetic on insects. He was so good, he could get decapitated abdomens of moths to lay eggs. Still, Wigglesworth is remembered as the "father of insect physiology," which—considering his foundational contributions not just to metamorphosis but also to respiration, digestion, water conservation, epidermal development, cuticle formation, and much more—seems right. And though he wasn't the one to apply his hormone work to insecticides, the bacterium *Wigglesworthia glossinidia*, which lives in the guts of tsetse flies, is named after him, somehow completing a circle. Today, the use of analogues of insect hormones is showing great promise: Just five grams of the nontoxic juvenile hormone derivative methoprene, for instance, will cover an acre of ground of mosquito-infested floodwater, and the derivative pyriproxyfen does wonders stunting the growth of *Anopheles*, the mosquito that serves as the vector of malaria. Both Darwin and Jean-Henri Fabre would have been tickled to learn of this innovation. The latter's rich biographical descriptions of the lives of insects, rather than in the clinical scientific mode, inspired Nabokov, Wigglesworth, and many others to become entomologists, and while he shunned the theory of evolution, Darwin nevertheless called him an "inimitable observer." As ever, Darwin's letters to him, quoted in the text, can be found with a quick search online at the UK University of Cambridge–based Darwin Correspondence Project. Darwin's own experience with the kissing bug is reported in his *The Voyage of the Beagle*, and the scholar who first suggested that his famous illnesses were due to Chagas disease was Saul Adler, in "Darwin's Illness," *Nature* 184 (1959): 1102–1103. Others have disputed this interpretation since, citing irritable bowel syndrome. Poor Darwin. Poor Wigglesworth sons. Poor kissing bugs.

6. "The Molecular Trinity": To reach Friday Harbor island, you drive seventy-five miles north from Seattle, and catch a sixty-five-minute ferry ride from Anacortes across Rosario Strait. If the sky is clear, Mount Baker with its snowcap will accompany your entire journey. As for Lynn and Jim, they still travel the world—immediately after I visited them, they were on their way to a three-week journey to India—but since 2007 they call the island their home. Looking back now at seventy years of a career in science, they are able to see more clearly which of their many studies have had an impact, and I thank them for pointing them out to me. For the original work done by Jim on clocks and PTTH "gating," see his and Lynn's "Neuroendocrine Control of Ecdysis in Silkmoths," *Science* 312 (1970): 1624–1626; "Physiology of Insect Rhythms. I. Circadian Organization of the Endocrine Events Underlying the Molting Cycle of Larval Tobacco Hornworms," *Journal of Experimental Biology* 57 (1972): 805–820; and "Physiology of Insect Rhythms. III. The Temporal Organization of the Endocrine Events Underlying Pupation of the Tobacco Hornworm," *Journal of Experimental Biology* 60 (1974): 371–382. For his work on "critical periods" relating to cuticle coloration, see his and Lynn and L. Safranek's "Hormonal Control of Cuticle Coloration in the Tobacco Hornworm, *Manduca sexta*: Basis of an Ultrasensitive Assay for Juvenile Hormone," *Journal of Insect Physiology* 19 (1973): 195–203, and also their "Temporal Patterns of Response to Ecdysone and Juvenile Hormone in the Epidermis of the Tobacco Hornworm, *Manduca sexta*," *Developmental Biology* 39 (1974): 247–262. A paper assessing the relative importance of external and internal cues for bringing about pupation is Jim W. Truman and Oliver S. Dominick, "The Physiology of Wandering Behavior in *Manduca sexta*. I. Temporal Organization and the Influence of the Internal and External Environments," *Journal of Experimental Biology* 110 (1984): 35–51. On Carroll Williams's and Fred Nijhout's work on "critical weight," see their "Control of Moulting and Metamorphosis in the Tobacco Hornworm, *Manduca sexta* (L.): Growth of the Last-Instar Larva and the Decision to Pupate," *Journal of Experimental Biology* 61 (1974): 481–491. For Lynn's work on pupal commitment, see her "Hormonal

Control of Insect Epidermal Cell Commitment in Vitro," *Nature* 259 (1976): 115–117.

I spent nearly a year in Jens Rolff's lab at the Freie University in Berlin. The lab is primarily an insect immunology lab searching for viable alternatives for antibiotics, but Jens has a deep interest in many other topics, including metamorphosis, and was kind enough to show me the ropes. One thing I learned is that trying to manipulate juvenile hormone by cutting out the *corpus allatum* is no walk in the park—allatectomies need practice, and I sincerely apologize to all the many beetles who paid the price! Another thing I learned was how precisely one needs to follow the life cycle of insects: To perform genetic manipulations in interrogating metamorphosis you have not only to be a careful molecular biologist but also a seasoned naturalist. All this helped me gain an even greater appreciation for Jim and Lynn's work.

On the working out of the role of *broad* with B. Zhou, K. Hiruma, and T. Shinoda, see Lynn's "Juvenile Hormone Prevents Ecdysteroid-Induced Expression of Broad Complex RNAs in the Epidermis of the Tobacco Hornworm, *Manduca sexta*," *Developmental Biology* 203 (1998): 233–244; and Lynn and Zhou's two papers, "Hormonal Regulation and Patterning of the Broad-Complex in the Epidermis and Wing Discs of the Tobacco Hornworm, *Manduca sexta*," *Developmental Biology* 231 (2001): 125–137; and "Broad-Complex Specifies Pupal Development and Mediates the 'Status Quo' Action of Juvenile Hormone on the Pupal-Adult Transformation in *Drosophila* and *Manduca*," *Development* 129 (2002): 2259–2269. On the perfection of the genetic allatectomy method, see L. M. Riddiford, J. W. Truman, C. K. Mirth, and Y. Shen, "A Role for Juvenile Hormone in the Prepupal Development of Drosophila melanogaster," *Development* 137 (2010): 1117–1126. Not everyone believes that the last word has been said about *chinmo*. As Xavier Bellés reported to me, there are currently three groups, one in Japan, another in the US, and his own group in Spain, working on the role of *chinmo* in hemimetabolan nymphs. It seems that in nymphs, *chinmo* also maintains the juvenile form (by reducing it in nymphs, the insect becomes an adult prematurely). If that's true, according to Bellés, *chinmo* would not be a specifying factor for the larva (in holometabolans), but for the larva and the nymph (hemimetabolans), meaning that the conclusions of Lynn and Jim's paper are not definitive and will be revised; new work on this matter is set to be published in 2025. Still, for now, Lynn and Jim's capstone paper putting the "molecular trinity" story together is J. W. Truman and L. M. Riddiford, "*Chinmo* Is the Larval Member of the Molecular Trinity That Directs *Drosophila* Metamorphosis," *Proceedings of the National Academy of Sciences* 119 (2022): e2201071119. The original speculative (and forgotten) paper by Williams and Kafatos on three possible master genes is "Theoretical Aspects of the Action of Juvenile Hormone," *Mitteilungen der Schweizerischn Entomologischen Gesellschaft* 44 (1971): 151–162; and Nancy Moran's paper on "adaptive decoupling" is "Adaptation and Constraint in the Complex Life-Cycle of Animals," *Annual Review of Ecology Evolution and Systematics* 25 (1994): 173–300. For the basic logic of master genes and gene regulatory networks, see E. H. Davidson and D. H. Erwin, "Gene Regulatory Networks and the Evolution of Animal Body Plans," *Science* 311 (2006): 796–800. And the symphony-like, and unstraightforward, way in which development works in the eye, and in mating behavior and sex, can be seen in L. M. Riddiford, J. W. Truman, and A. Nern, "Juvenile Hormone Reveals Mosaic Developmental Programs in the Metamorphosing Optic Lobe of *Drosophila melanogaster*," *Biology Open* 7 (2018): bio034026; and in J. Bilen, J. Atallah, R. Azanchi, J. Levine, and L. M. Riddiford, "Regulation of Onset of Female Mating and Sex Pheromone Production by Juvenile Hormone in *Drosophila melanogaster*," *Proceedings of the National Academy of Sciences* 110 (2013): 18321–18326, respectively.

I am grateful to Mary and Fred Nijhout who generously answered my questions about this period in the lives and research of Jim and Lynn, as well as to Xavier Bellés for his sensitive intelligence and warmth. For those interested in the bigger picture in biology that allowed Lynn and Jim's work, there is still no better source than the wonderful book by Horace Freeland Judson, *The Eighth Day of Creation* (Simon and Schuster, 1979), which provides a detailed history of the molecular revolution. And for those trying to understand why genes are more complex than the initial classical definition suggested, read Kostas Kampourakis, *Understanding Genes* (Cambridge University Press, 2021). Finally, this period in Lynn's professional life saw her and her students studying other questions. For an example of a beautiful piece of work outside the realm of metamorphosis, see Lynn's paper with her graduate student R. G. Vogt that uncovered how moths smell, "Pheromone Binding and Inactivation by Moth Antennae," *Nature* 2293 (1981): 161–163. A post-doc working with Lynn at the time went on to become a leading host-parasite researcher; see L. M. Riddiford and B. A. Webb, "Nancy E. Beckage (1950–2012): Pioneer in Insect Host-Parasite Interactions," *Annual Review of Entomology* 59 (2014): 1–12. Maria Sibylla Merian would have been blown away by Nancy Beckage's work, but perhaps most of all by the discovery that parasitoids manipulate the hormones, and therefore the genes, of their hosts.

7. "Hopeful Monsters": Frank Ryan's book *The Mystery of Metamorphosis: A Scientific Detective Story* (Chelsea Green, 2011) is responsible for the hard-won research on Donald Williamson provided in this section. The book is forwarded, unsurprisingly, by Dorion Sagan and his mother Lynn Margulis. It was praised by some, and attacked by others, including Josh Trapani, somewhat unfairly in my view, in his review "This Look at a Fascinating Subject in Evolution Takes a Wrong Path," in the *Washington Independent*, May 30, 2011. Williamson's own major writings can be found first in his initial paper, "Incongruous Larvae and the Origin of Some Invertebrate Life-Histories," *Progress in Oceanography* 19 (1987): 87–116, then in his two subsequent books, *Larvae and Evolution: Toward a New Zoology* (1992); and *The Origin of Larvae* (2003); and later in the paper "Hybridization in the Evolution of Animal Form and Life-Cycle," *Zoological Journal of the Linnean Society* 148 (2006): 585–602. Williamson's paper "Caterpillars Evolved from Onychophorans by Hybridogenesis," *Proceedings of the National Academy of Sciences*, was published online on August 28, 2009, but the journal's editor-in-chief, Randy Schekman, decided not to publish it in hard copy, and it was strongly rebutted in that same journal by Michael W. Hart and Richard K. Grosberg in "Caterpillars Did Not Evolve from Onychophorans by Hybridogenesis," on October 30 of the same year. (Michael Hart's earlier refutation of Williamson's claims was published in his article "Testing Cold Fusion: Maternity in a Tunicate x Sea Urchin Hybrid Determined from DNA Comparisons," *Evolution* 50 (1996): 1713–1718.) For more on the brilliant and tragically young-dying Balfour, see Brian K. Hall, "Francis Maitland Balfour (1851–1881): A Founder of Evolutionary Embryology," *Journal of Experimental Zoology B. Molecular Developmental Evolution* 229 (2003): 3–8. The quote from Lynn Margulis cited in the text appears in Elie Dolgin, "Row at US Journal Widens," *Nature*, October 9, 2009. For more on her ideas, see Lynn Margulis and Dorion Sagan, *Acquiring Genomes: A Theory of the Origin of Species* (Basic Books, 2003). And for wider, philosophical discussions of this fascinating topic in evolution, see Alfred I. Tauber's edited book *Organism and the Origins of Self* (Boston Studies in the Philosophy and History of Science, 1991).

8. "Origin": The evolution of insect metamorphosis is a complicated matter, with experts who argue one thing strongly opposed by experts who argue another. That a velvet worm climbed on land once has been challenged by evidence pointing both to multiple events of terrestrialization by different arthropod groups, some or all of which already had hard parts and even legs; see for example the evidence presented in the essay "Water-to-Land-Transitions" in *Arthropod Biology and Evolution*, edited by Allessandro Minelli, Geoffrey Boxshall, and Giuseppe Fusco (Springer, 2013). But before jumping into the fray, it's worthwhile reading a history of the subject: D. F. Erezylmaz, "Imperfect Eggs and Oviform Nymphs: A History of Ideas About the Origin of Insect Metamorphosis," *Integrated Computational Biology* 46 (2006): 795–807. Lovers of Aristotle will enjoy S. E. Reynolds, "Cooking Up the Perfect Insect: Aristotle's Transformational Idea About the Complete Metamorphosis of Insects,"

Philosophical Transactions of the Royal Society of London B 374 (2019): 0074. And the Berlese-Imms versus Hinton hypotheses can be read in the original here: Antonio Berlese, "Intorno Alle Metamorfosi Delli Insetti," *Redia* 9 (1913): 121–136; and H. E. Hinton, "The Origin and Function of the Pupal State," *Proceedings of the Royal Entomological Society of London A* 38 (1963): 77–85.

Lynn and Jim's first foray into the evolutionary realm appeared as "The Origins of Insect Metamorphosis," *Nature* 401 (1999): 447–452. They later tweaked and updated their approach in "The Evolution of Insect Metamorphosis: A Developmental and Endocrine View," *Philosophical Transactions of the Royal Society B* 374 (2019): 20190070 (and see also Marek Jindra, "Where Does the Pupa Come From? The Timing of Juvenile Hormone Signaling Supports Homology Between Stages of Hemimetabolous and Holometabolous Insects," *Philosophical Transactions of the Royal Society of London B* 374 [2019]: 20190064). Their molecular trinity paper, "*Chinmo* Is the Larval Member of the Molecular Trinity That Directs *Drosophila* Metamorphosis," *Proceedings of the National Academy of Sciences* 119 (2022): e2201071119; and their firebrat article (written with Czech colleagues), "The Embryonic Role of Juvenile Hormone in the Firebrat, *Thermobia domestica*, Reveals Its Function Before Its Involvement in Metamorphosis," *eLife* (2023): 12:RP92643, strengthen Lynn and Jim's resolve that *chinmo-broad-E93* is a kind of temporal Hox gene system. Hox genes actually got their name from the term "homeotic transformations," which William Bateson had coined to describe gene mutations that cause the development of one body part or structure to be transformed into another body part or structure. Today, we know that Hox genes are a family of homeotic transcription factors that play a key role in controlling the body plan along the anterior-posterior axis. On the history of how Hox genes were discovered and how they work, a beautiful book by Sean B. Carroll, himself an important player in this field, is *Endless Forms Most Beautiful: The New Science of Evo-Devo and the Making of the Animal Kingdom* (Quercus Publishing, 2011). For a more technical account of the central role played by changes to regulatory sequences in the evolution of form, see the textbook of another pioneering geneticist, Eric H. Davidson, *Genomic Regulatory Systems: In Development and Evolution* (Academic Press, 2001). Those interested in diving deeper into the historical origins and philosophical ramifications of evo-devo in humans should read Ron Amundson's *The Changing Role of the Human Embryo in Evolutionary Thought: Roots of Evo-Devo* (Cambridge Studies in Philosophy and Biology, Cambridge University Press, 2007); and a wonderfully rich scientific book about the subject is Mary Jane West-Eberhard's *Developmental Plasticity and Evolution* (Oxford University Press, 2003).

I am grateful to the generosity and wisdom of Stuart Reynolds, who helped me better understand the different strands of thinking about the evolution of metamorphosis, especially how the birth of the larva and the birth of the pupa are connected. However complete metamorphosis got started, why it became so successful is a separate question. The gut microbiota replacement theory is presented in Paul R. Johnston, Véronique Paris, and Jens Rolff, "Immune Gene Regulation in the Gut During Metamorphosis in a Halo- Versus Hemimetabolous Insect," *Philosophical Transactions of the Royal Society of London B* 374 (2019): 20190073; as well as Tobin J. Hammer and Nancy A. Moran, "Links Between Metamorphosis and Symbiosis in Holometabolous Insects," *Philosophical Transactions of the Royal Society of London B* 374 (2019): 20190068. The original idea of breaking up genetic correlations between different life-cycle stages can be read in Vincent B. Wigglesworth, *The Physiology of Insect Metamorphosis* (Cambridge University Press, 1954). And the related idea of decoupling growth and differentiation is presented in Jens Rolff, Paul R. Johnston, and Stuart Reynolds, "Complete Metamorphosis of Insects," *Philosophical Transactions of the Royal Society of London B* 374 (2019): 20190063, as well as the more recent Christine Manthey, C. Jessica E. Metcalf, Michael T. Monagham, Ulrich Karl Steiner, and Jens Rolff, "Rapid Growth and the Evolution of Complete Metamorphosis in Insects," *Proceedings of the National Academy of Science* 121, no. 3 (2024): e2402980121, in which the authors show that holometabolous insects' pre-adult growth rates are faster than hemimetabolous ones. The evo-devo notion of holometaboly contributing to a creature's evolvability is explained in M. Kirschner and J. Gebhart, "Evolvability," *Proceedings of the National Academy of Sciences* 95 (1998): 8420–8427. As for complete metamorphosis reducing competition between generations by adapting juveniles and adults to different foods and different environments, everyone seems to agree on that (for a clear exposition, see R. E. Snodgrass, *Insect Metamorphosis*, Smithsonian Miscellaneous Collections, 22, no. 9 [1954]). It is true, as Dino McMahon has shown, that some insects stave off the costs of becoming adults, foregoing complete metamorphosis like axolotls, but these seem to be very rare exceptions that prove the rule, and only ever females besides; see his "Why Grow Up? A Perspective on Insect Strategies to Avoid Metamorphosis," *Ecological Entomology* 41, no. 5 (2016): 505–515. Finally, for those intrigued by the flight hypothesis explaining the birth of hemimetaboly, read this cool paper on wing evolution in insects: Xavier Bellés, "The Innovation of the Final Moult and the Origin of Insect Metamorphosis," *Philosophical Transactions of the Royal Society of London B* 374 (1783): 20180415. The author, a leading metamorphosis expert alongside Lynn and Jim who played an important role in finding *E93*, and a close friend of theirs, argues that it was the challenge of flight that got incomplete metamorphosis (and by extension complete metamorphosis) rolling.

9. "Thyroxine": Thyroxine was first isolated by Edward Calvin Kendall in 1915, more than thirty years after the death of Charles Darwin. Since it is partially composed of iodine, iodine deficiency leads in humans to a disease called goiter in which the thyroid gland is enlarged, sometimes making it hard to breathe and often adversely impacting weight gain. Levothyroxine is a manufactured form of thyroxine used to treat the problem; in 2018 it was the second most commonly prescribed medicine in the United States. But all this refers to normal physiological maintenance. For the role of the thyroid system during the special periods of adolescence and fetal development, respectively, see Maria Christina Vigone, L. Stroppa, and

G. Chiumello, "Thyroid Function and Puberty," *Journal of Pediatric Endocrinology and Metabolism* 16 (2003): 253–257, and Juan Bernal, "Thyroid Hormone Receptors in Brain Development and Function," *Nature Reviews Endocrinology* 3 (2007): 249–259. Thyroid hormones, in other words, are important to humans. The original paper that reported thyroid action in mammalian fetal brain development is A. L. Dowling, G. U. Martz, J. L. Leonard, and R. T. Zoeller, "Acute Changes in Maternal Thyroid Hormone Induce Rapid and Transient Changes in Gene Expression in Fetal Rat Brain," *Journal of Neuroscience* 20, no. 6 (2000): 2255–2265. Much remains to be learned about the role of thyroid hormones in human brain development, but there's also been major progress. To get up to speed, see Barbara K. Stepien and Weiland B. Huttner, "Transport, Metabolism and Function of Thyroid Hormones in the Developing Mammalian Brain," *Frontiers in Endocrinology* 10 (2019): article 209. In constructing this section's argument, I drew upon pages 213–219 and 234–242 in Ryan's aforementioned *The Mystery of Metamorphosis*, where the seminal work of R. Thomas Zoeller and his collaborators on thyroid hormones in the mammalian brain is explicated in greater detail.

As important as thyroid hormones are for us, they didn't originate with humans. For a brief history, see J. R. Tata, "Getting Hooked on Thyroid Hormone Action: A Semi-autobiographical Account," *Journal of Bioscience* 33 (2008): 653–667, as well as a good explanation of the workings of thyroid hormones in vertebrates in Vincent Laudet, "The Origins and Evolution of Vertebrate Metamorphosis," *Current Biology* 21 (2011): R726–R737. Also informative are these two articles on the advantages and disadvantages of hormone-controlled metamorphic plasticity: E. M. Callery and R. P. Elinson, "Thyroid Hormone-Dependent Metamorphosis in a Direct Developing Frog," *Proceedings of the National Academy of Sciences* 97 (2000): 2615–2620; and K. M. Warkentin, "Plasticity of Hatching in Amphibians: Evolution, Trade-Offs, Cues and Mechanisms," *Integrative and Comparative Biology* 51, no. 1 (2011): 111–127. To understand how the capturing of iodine by primitive marine creatures got this all started, see Susan J. Crockford, "Evolutionary Roots of Iodine and

Thyroid Hormones in Cell-Cell Signaling," *Integrative and Comparative Biology* 49, no. 2 (2009): 155–166. On thyroxine in ascidian larvae, and the endostyle, see Paolo D'Agiti and Matteo Cammarata, "Comparative Analysis of Thyroxine Distribution in Ascidian Larvae," *Cell and Tissue Research* 323 (2006): 529–535. More on the evolutionary origins of this system can be found in Elias Taylor and Andreas Heyland, "Evolution of Thyroid Hormone Signaling in Animals: Non-Genomic and Genomic Modes of Action," *Molecular and Cellular Endocrinology* 459 (2017): 14–20. In the same issue of the journal, see Richard Z. Manzon and Lori A. Manzon, "Lamprey Metamorphosis: Thyroid Hormone Signaling in a Basal Vertebrate," pp. 28–42, to clarify the link to the sucking lamprey. For more on the research potential of ascidians, see Joseph C. Corbo, Anna Di Gregorio, and Michael Levine, "The Ascidian as a Model Organism in Developmental and Evolutionary Biology," *Cell* 106, no. 6 (2001): 535–538. As for the relationship between sea creatures and insects, that's a little less straightforward: After all, in marine invertebrates like the sea squirt and starfish, the larva stage is short, while the adult stage is long. In insects, it's the opposite: Think of an achingly slow thirteen- or seventeen-year developing cicada in the soil, or a mouthless mayfly on its final molt. Besides, all insects belong to one class, the mandibulates, whereas marine invertebrates are strewn among many phyla. And yet, despite the differences, there are commonalities beyond denial. For one, the imaginal discs of metamorphosing insects and the pluripotent cells that line the stomachs of marine invertebrates both represent the very same reservoir. Whether they evolved through hybridization as Williamson thought, or simply through gradual mutation on the Darwinian model, they were both "set aside" stems cells. They were there—at a particular point in the organism's life cycle—to climax a new developmental program, and so they were invented time and again. The family of hormone-induced transcription factors that are common to both insects and vertebrates is a further example of convergent evolution. On the differences and similarities between thyroid and ecdysone-based hormonal signaling, see Thomas Flatt, Leonid L. Moroz, Marc Tatar, and Andreas

Heyland, "Comparing Thyroid and Insect Hormone Signaling," *Integrative and Comparative Biology* 46, no. 6 (2006): 777–794; and in the same issue, Noah Ollikainen, Charlie Chandsawangbhuwana, and Michael E. Baker, "Evolution of the Thyroid Hormone, Retinoic Acid, Ecdysone, and Liver X Receptors," pp. 815–826. I'll say it again: There is grandeur in this view of life.

But alongside the grandeur there is also a terrible irony. Rather than celebrating our origins in the oceans and protecting our far-off cousins, we humans are leading to their demise. One recent study has shown that increased temperature and pesticides disrupt the thyroid pathways in Manini, the convict surgeonfish, so called for their bold black stripes on a yellowish background, like a convict in jail. This small and widespread fish uses thyroid hormone pathways both for metamorphosis and to direct behaviors that allow it to escape from predators; see Marc Besson et al., "Anthropogenic Stressors Impact Fish Sensory Development and Survival via Thyroid Disruption," *Nature Communications* 11 (2020): article 3614. If we don't take care to safeguard its environment, and the environment of millions of other species in the oceans, our negligence will eventually lead to their extinction.

10. "Rosetta Stone": The first paper Jim Truman published about the insect nervous system appeared exactly fifty years ago: Hugh M. Taylor and Jim W. Truman, "Metamorphosis of the Abdominal Ganglia of the Tobacco Hornworm, *Manduca sexta*," *Journal of Comparative Physiology* 90 (1974): 367–388. His latest paper on the subject is James W. Truman, Jacquelyn Price, Rosa L. Miyares, and Tzumin Lee, "Metamorphosis of Memory Circuits in *Drosophila* Reveals a Strategy for Evolving a Larval Brain," *eLife* 12 (2023): e80594. In between, Jim was influenced by many studies and scholars, in particular Michael Bate's first map of neuroblasts in the ventral nervous system of locusts, published in 1976, which showed a repeating array of sixty-one neuroblasts (thirty pairs, one unpaired) that generated the neurons in each segmental ganglion; Corey Goodman's later work, in collaboration with Bate, showing that each neuroblast produced a characteristic set of neurons and developed as an autonomous module; and John Thomas's work, in collaboration with the two others, showing how the rules of insect nervous system development have been conserved in evolution. The two most relevant papers are C. M. Bate, "Pioneer Neurons in an Insect Embryo," *Nature* 260 (1976): 54–56; and John B. Thomas, Michael J. Bastiani, Michael Bate, and Corey S. Goodman, "From Grasshopper to *Drosophila*: A Common Plan for Neural Development," *Nature* 310 (1984): 203–207. A crucial step on the way to Jim's 2023 *eLife* paper, dubbed "The Paper" by many colleagues, was showing that the hemilineages derived from initial neuroblasts in the fly nervous system were developmentally and behaviorally equivalent: Robin M. Harris, Barret D. Pfeiffer, Gerald M. Rubin, and James W. Truman, "Neuron Hemilineages Provide the Functional Ground Plan for the *Drosophila* Ventral Nervous System," *eLife* 4 (2015): e4493. The milestone "connectome" paper that ultimately came out of Janelia was Zheng et al., "A Complete Electron Microscopy Volume of the Brain of Adult *Drosophila melanogaster*," *Cell* 174, no. 3 (2018): 730–743. But the question of what memories of a larval brain make it through metamorphosis into an adult brain continues to be studied and clearly depends on the species and its degree of refashioning of its brain as it goes through the life cycle. Early evidence that memories can be retained in beetles can be found in Thomas M. Alloway, "Retention of Learning Through Metamorphosis in the Grain Beetle," *American Zoologist* 12 (1972): 471–477. A more recent study of lepidoptera is Douglas J. Blackinston, Elena Silva Casey, and Martha R. Weiss, "Retention of Memory Through Metamorphosis: Can a Moth Remember What It Learned as a Caterpillar?," *PLOS ONE* 3, no. 3 (2008): e1736. L. Bégué, N. Tschirren, M. Peignier, et al. take things one step further, looking at amphibian personality in "Behavioural Consistency Across Metamorphosis in a Neotropical Poison Frog," *Evolutionary Ecology* 38 (2024): 157–174. (For a broader survey of the fascinating subject of animal personality, see Doreen Cabrera, Joshua R. Nilsson, and Blaine D. Griffin, "The Development of Animal Personality Across Ontogeny: A Cross-Species View," *Animal Behavior* 173 (2021): 137–144.) I thank Gerit Linneweber, Emmy Noether fellow and group leader of "Origins of Individuality" in the department of biology at the Freie University, for a wonderful discussion about the relationship between nervous system architecture and individuality. Gerit and his group are doing fascinating work on how nonheritable stochastic changes to brain wiring in flies result in unique specimens with unique behaviors (particularly with respect to orienting toward visual objects). See their publication in *Science* from 2020, "A Neurodevelopmental Origin of Behavioral Individuality in the *Drosophila* Visual System." Finally, Jim Truman told me of an experiment in which he starved *Drosophila* larvae of protein. The adults emerging from these larvae had 20 percent more neurons on their mushroom bodies, which figure in olfactory learning. Though he hasn't yet tested the learning abilities of these flies, it is possible that experiencing aversive conditions when young might produce "smarter" flies, better able to deal with aversive conditions as adults. Humans think of the "self" in terms of conscious awareness, but experience also affects subconscious parts of the nervous system. Since whatever a "self" may be must include both conscious and subconscious elements, do we ever really know our complete self? Relatedly, the Kant quote about natural science not being able to discover the inside of things comes from his *Prolegomena* from 1783.

11. "Sol": What the self is, and how to come to terms with it, are questions that humans have contemplated for millennia, in philosophy, science, and art. Plutarch famously discusses the Argo in his *The Life of Theseus*, and John Locke spoke of the prince and the cobbler in Book 2 of his *An Essay Concerning Human Understanding*, published in 1690. In Eastern cultures, Buddhism preaches that there is no self, that the self is an illusion. Hinduism, to the contrary, teaches that the only thing that exists is the self—the absolute realm of truth where everything is one. The swami Achyutamrita Chaitanya and Dr. Bhavani Rao, with whom I had long conversations during the writing of this book in Amma's ashram in Amritapuri, Kerala, India, taught me that in their tradition, finding yourself is finding the one self of everything; therefore, love yourself so that you can love others. For more on the concept of the self from a Western sociological and political standpoint, see Anthony Elliott, *Concepts of the Self*, 4th edition (Polity, 2020), and for a survey that includes philosophical, scientific, and religious perspectives, see *The Oxford Handbook of the Self* (Oxford, 2013). As for the *New Yorker* article, "Are You the Same Person You Used to Be?," it was authored by Joshua Rothman and appeared on October 3, 2022, and I have him to thank for the references to the Dunedin Study, and Galen Strawson. A recent piece by Strawson in the *Dublin Review of Books* is called "Just Live," and provides a good representative argument for his "episodic" approach to life. Finally, not everyone agrees with the ideas of Derek Parfit, who argued until his death in 2017 that while identity can't be entirely reduced to just brains and bodies, identity doesn't matter for survival. For Parfit, not believing in the importance (or existence) of a separate self was liberating: As differences between people shrink, he could care less about himself and more about others. See his *Reasons and Persons* for the full argument. Others find a strong sense of self crucial for meaning, as well as mental health.

The birth of our third child completed our family. These days, Sol is five, as wild and wise as she was born. This book is dedicated to her.

INDEX

INDEX

N

Credit: *Yaal Herman*

Oren Harman is senior research fellow at the Van Leer Jerusalem Institute and teaches at the Graduate Program in Science, Technology, and Society at Bar-Ilan University. His books include *Evolutions* and *The Price of Altruism*, which won the Los Angeles Times Book Prize. He lives in Berlin and Jerusalem.